Electric Energy Generation

Electric Energy Generation

Economics, Reliability, and Rates

Joseph Vardi
and Benjamin Avi-Itzhak

The MIT Press
Cambridge, Massachusetts,
and London, England

This book was set in Univers and Times Roman by Komiyama Printing Company and printed and bound by The Alpine Press Inc. in the United States of America.

Library of Congress Cataloging in Publication Data

Vardi, Joseph.
Electric energy generation.

Based on the first author's thesis, Technion-Israel Institute of Technology.
Bibliography: p.
Includes index.
1. Electric utilities—Management. 2. Electric utilities—Management—Mathematical models.
I. Avi-Itzhak, Benjamin, 1933– joint author.
II. Title.
HD9685.A2V37 363.6′2 80–20462
ISBN 0–262–22024–5

Contents

Preface

The increasing importance of energy and its proper supply, and the magnitude of the expenditures associated with it, created an urgent need to sharpen and develop the analytical tools available to policymakers. In order to employ these tools effectively a better understanding is required of their potential usages, how they work, and what can be expected from them.

Within the energy sector the electric power subsector, which is the largest single consumer of hydrocarbons, plays a central role by providing a most important means for large-scale conversion, transportation, and distribution of energy. This subsector also presents the largest single requirement for capital expenditures.

During the last 15 years considerable advances have been achieved in the development of analytical tools for electric power planning and policy issues, with rate making, reliability analysis, and forecasting of system behavior attracting much interest. In the field of rate making the pioneering works and concepts of the 1950s and the 1960s of, among others, Electricity of France Company in France, Turvey (33) in England and Kahn (20) in the United States, using the marginal cost approach, gained wide acceptance and was followed by many refinements and improvements. In the same period the introduction of operations research and computer application to electric power system planning generated interest in better tools for evaluating electric energy generating systems. In 1967 Jemolle (3) in Belgium layed the foundation for what is known today as probabilistic simulation. This method was later followed by the Australian Booth (9) and the Americans Jenkins and Joy (15), who implemented this approach for the International Atomic Energy Agency in a set of programs called WIEN or WASP. This method successfully introduced reliability concepts into system evaluation techniques for the first time, thus overcoming a major deficiency of earlier methods. The industry's interest in reliability is as old as the industry itself, but considerable interest and exposure, as well as analytical treatment, were attained only after the major blackout in New York in 1967.

The need for a book addressing issues of economics, reliability, and rates in electric energy systems planning has been felt for quite some time. Presently there are in existence several good books dealing with electric energy systems planning; each, however, specializes in only some of the issues. The purpose of this book is twofold: first, to introduce the interested reader to the terminology, concepts, and main issues of the modern electric power generating industry; and second, to provide an up-to-date comprehensive discussion of one unified, analytical approach applicable to many of the different issues involved in the system analyses of electric energy generation.

Every book presents its authors with the problem of what should be included and what omitted. In our case it was decided that the interested reader will benefit more from a full discussion of a wide range of possible applications of one central method than from a great number of different methods. By this approach the unexperienced reader will become acquainted with all the relevant basic concepts and at the same time master a helpful analytical tool. The experienced reader will be provided with a discussion of an important technique that very rapidly is gaining wide acceptance in the industry.

The book is aimed at persons interested in energy economics, system planning, energy and pricing regulation, rate making, system reliability, and decision making. It is also intended for use as a textbook or reference book in electric energy economics and system planning of electric energy generating systems as a part of a wider curriculum in energy, economics, operations research, and electric power engineering. In view of the diverse backgrounds and interests of the intended readers the book requires only elementary knowledge of the fundamentals of probability and calculus. It is believed that the nonmathematician will find no difficulty in following the discussion.

The material in this work is an extension of a doctoral dissertation by the first author written under the guidance of the second author at the Technion–Israel Institute of Technology; it is based in part on a number of recent publications coauthored by Dr. Jacob Zahavi of Tel-Aviv University, to whom the authors are grateful. The book was written while Dr. J. Vardi was serving as the Director General of the Israel Ministry of Energy and Professor B. Avi-Itzhak was visiting at Stanford University. The authors are indebted to the Systems Optimization Laboratory of Stanford University for generous secretarial and typing help.

Haifa, January, 1980

Electric Energy Generation

1

Introduction

1.1 The Electric Power Industry

Electric power constitutes a major portion of total energy consumption of most countries. In 1976, for instance, total electric generation in the U.S. amounted to 2.1 trillion kWh (11). Fuel consumption by generators of all types (coal, oil, gas, nuclear, etc.) accounted for approximately 30% of the total U.S. primary energy consumption. All indicators show that demand for electric power will continue to grow at a relatively high rate, increasing electricity's share of fuel consumption. To meet the expected increase in demand, and ensure reliable and uninterrupted supply of power, vast investments are needed for the construction of generating facilities and transmission and distribution networks. By and large, the electric power industry is the largest single investor in the U.S. economy; this tendency will become even more dominant in the future as a result of switching to new technologies and automation. Investment of investor-owned electric utilities (which constitute approximately 80% of the entire power industry in the U.S.) amounted to $16.9 billion in 1976 alone (11). The operating costs, mainly fuel costs, of satisfying the demand for power exhibit a similar pattern; for example, the operating cost incurred by the investor-owned power utilities amounted to $37.6 billion in 1976 (11). This figure is most likely to increase in future years because of high inflation rates, rising fuel prices, and expected increase in demand.

1.2 The Scope of the Book

The magnitude and complexity of the electric power industry, and its increasing role in our lives, have significantly complicated the planning process. The first part of the book reviews the traditional basic concepts and issues in the analysis of electric power generating systems, including the different approaches of system analysis, cost calculations, reliability analysis, and marginal cost approaches to rate making. The complexity of the decision problems has led to the development of a host of diversified techniques ranging from simple graphical devices to sophisticated simulation and mathematical programming models, each of which pertains to a certain class of problems or a different type of application. This diversity of the decision models often results in a fragmentation of the decision-making process, increased cost of making the decisions, lack of insight into the problem, and decreased level of accuracy.

The second part of the book presents a unified methodological approach capable of supporting a wide range of system analysis needs in the electric power generation field. The approach is based on the recently developed equivalent load–duration curve representation. This part of the book formulates the unified approach and extends its application to three major areas of electric power system analysis: reliability analysis, operating cost forecasting, and rate making. The methodology is especially equipped to deal with the probabilistic behavior of the electric generating system.

The importance of incorporating into the model the probabilistic nature of the system cannot be overstated. The probabilistic element is a major problem in any comprehensive model dealing with the analysis of an electric power generating system. Incorporation of the stochastic behavior of the system into the model makes it possible mainly to capture the effects of forced outages. Due to equipment failures, generating units are idle for various periods, affecting the requirements for installed capacity to meet given reliability or grade-of-service criteria. Outages also affect the fuel consumption of the system since different units consume different quantities of fuel. In addition, when an analysis of the economics of peak units is performed, it is essential to find as exactly as possible how many hours each particular peak unit will work. The number of hours is affected less by the forced outages rate of the particular unit than by the forced outages rates of all the other units of the system.

Most models, including those described in the first part of the book, try to overcome these probabilistic effects either by ignoring the whole issue or by correcting it through insertion of some penalty for the reduced availability of the units. While the latter approach does not greatly distort the calculations related to the base-load units, the higher the unit is on the merit order, the higher the deviation from real life, and the approach is totally unacceptable when peak units are concerned. One conventional remedy for overcoming these inaccuracies is to simulate the whole system. Simulation is a powerful tool. However, one can simulate only a limited number of alternatives at an acceptable expense.

The equivalent load–duration curve (ELDC) is a relatively new tool designed to overcome many of these difficulties. The ELDC is a convolution of the consumer load, the "demand" of capacity for maintenance, and the "demand" of capacity for failures. Maintenance and forced outages are regarded as another kind of demand for capacity with zero consumption of fuel. This approach was first introduced by Baleriaux et al. (3).

The main advantage of the convolution approach is that it incorporates into the ELDC outage probabilities, maintenance requirements, and random deviations of customers' load. Hence any change in the characteristics of the system will also affect the ELDC as each such change contributes to different customers' demand or to different demand for maintenance and forced outages. This property of the ELDC yields a tool which has the potential of being most powerful and versatile for electric power system analysis, and is exploited here to derive a host of analyses by studying the impact of parameters of the system through their reflection on the ELDC. The importance of these analyses to decision making is also demonstrated. For instance, based on the observation that any excess of combined demand over the installed capacity results in a shortage of power supply, the ELDC lends itself rather easily to reliability analysis, as has already been recognized by Joy and Jenkins (19) with regard to calculating the loss-of-load probability and various sensitivity analyses. The ability to perform conveniently reliability analysis is an important feature in decisions regarding reserve requirements. The ELDC has been successfully applied to the calculation of the expected operating cost of the system under a merit order loading, providing improved estimates of the expected generation requirements for each unit in the system. This book presents an ELDC-based method for calculating the operating cost under the more realistic, yet more complex, assumption of incremental cost loading.

A major advantage of the ELDC as an aid in decision making is the fact that it has been formulated in an explicit analytical manner. Such a formulation will undoubtedly open up more potential applications of the ELDC and at the same time provide better insight into the system and its behavior. Developing a tool to analyze the system by computation rather than by simulation enables one to compute rigorously many important parameters of the system in an easy way. Various applications of these methods are described, including finding

- the expected number of hours that each individual unit will work;
- the expected fuel consumption of each individual unit for the case of incremental cost loading and the fuel budget of the system;
- the effect of adding a unit on the reliability of the system;
- the effect of the size of a unit on the reliability of the system;
- the trade-off between better reliability of a unit versus the installed capacity required in a system for meeting a given criterion of system reliability;
- the loss-of-load probability;
- the distribution of capacity surplus;

- the distribution capacity deficiency;
- the expected amount of energy which cannot be supplied;
- how different customers should share energy and capacity costs.

I

Traditional Concepts and Issues in Evaluating Electric Power Generating Systems

2

Major Planning and Analysis Issues in the Field of Electric Power Generation

2.1 Planning Issues

"Planning is basically estimating and evaluating, whether it be for events in the immediate future, the middle range (5 to 7 years ahead) or the far future (15 to 30 years ahead)" (4). The major decision issues in the electric power generation field that fall under this definition, and are treated in this book along with the basic elements involved in the decision process, are the following.

Reliability Analysis

Reliability has become one of the most important design criteria of electric power systems. It is a most complex quantity, as it depends upon many factors, including unit size, unit type, reserve margin, load shape, forced outages rate, existence of interconnection, and the number of units in the system.

Reserve Planning

In order to maintain the desired level of reliability and ensure against excessive shortages, an extra generating capacity above the peak demand, called *reserve*, must be maintained. The question of what size of reserve to maintain in the system is one of the important decision problems of electric power companies, due to massive investments needed.

Operating Cost Analysis

Costs analysis, in which operating cost predictions play a major role, is important to almost any planning problem. Several particular areas of applications are described in (2).

Generation Planning and Costs

Since the goal of any power company is to supply power at least cost while ensuring a high level of reliability, operating cost and capital cost analyses are major elements in planning for the capacity expansion of power systems. Estimation of costs for up to 10 to 30 years is normally required for generating planning.

Financial Planning and Operating Costs

Financial planning is concerned with determining the ability of the power company to meet its expenses and analyze its need for capital. An important aspect of financial planning is cash-flow analysis, which also indicates how dependent the power company is on external capital markets. While production expense is only one item considered in financial planning, it is nevertheless a most important one, and its magnitude makes it necessary to calculate the operating cost with a high level of accuracy.

Fuel Budgeting

Fuel budgeting provides the power company with information on its need for fuel of all types in future years, thereby enabling the company to take the necessary steps to ensure an undisrupted supply of fuel. Planning horizons of 1 to 3 years are usually considered in this regard.

Rate Structure Analysis

The term *rate structure analysis* (or *tariff policy*, as is used in most other countries) refers to the task of determining the price of electricity. From the company's point of view, the rate structure is extremely important because it is the means by which the utility receives the revenues necessary to support its operations. From a national point of view, pricing of electricity is an increasingly important element in the allocation of energy resources. It is therefore essential to devise a price structure which will reflect the true value of electricity to the company and to society.

2.2 Analysis Methods

Comprehensive Methods versus Marginal Cost Comparison Methods

In accordance with the basic difference between short- and long-term planning, two basic approached have been devised to tackle the analysis of power systems: a *marginal cost comparison approach* for short-term planning; and a *comprehensive*, or *global*, *approach* for medium- and long-range planning. Global mathematical programming models (1) have been used in order to generate investment alternatives for power systems in the short and the long run, with digital simulation usually advocated as a means of further analyzing the results. In a sense, the optimization models are conceived of as "global" models, providing approximate investment policies, which are then further refined using "marginal" models. As Anderson (1) puts it, "Global models can only give approximate answers in most practical situations. The reason is that the practical details of the alternative programs, particularly of the individual projects in the programs, are too numerous to be handled in one computer run. Having obtained approximate solutions from the global models, we then turn to marginal analysis using situation models to focus on the fine details of individual project selection and design" (pp. 269–270).

While simulation models are undoubtedly very powerful tools for analyzing performance of a given electric power system, they are inflexible for marginal analysis, as each change in the data or the parameters of the system makes it necessary to rerun the simulation model under the new conditions. It is also quite difficult to incorporate simulation models into optimization models, and in many cases they fail to provide us with the required insight into the problem, as they generate numerical, rather than analytical, results.

Sensitivity Analysis and Marginal Methods

Sensitivity analysis is the task of determining the impact of perturbing the parameters of the system on its performance measures. The high variability of data, resulting from uncertainties, fluctuation of data over time, and long planning horizons, makes such an analysis particularly attractive for electric power system planning. In addition, sensitivity analysis serves to provide further insight into the relationships between the various elements of the system.

One aspect of sensitivity analysis is trade-off analysis, sometimes referred to as the marginal or break-even approach. Due to the conflicting nature of various parameters in electric power systems, it is often necessary to supplement planning decisions by trade-off analysis in order to balance the conflicting interests one against the other. In essence, trade-off analysis is the task of determining the change in a system parameter, such as the unit's size, required to compensate for the change in the system performance resulting from perturbing any other parameter of the system, say the unit's forced outages rate. By definition, trade-off analysis is a sensitivity analysis problem, though definitely of a more complex nature.

Marginal methods are used for detailed analysis of a given alternative investment, or for comparing a small number of investment programs, usually by employing economic accounting-type techniques or simulation. If a least-cost short-term investment is to be found by marginal cost comparison, it will be done by substitution of units which satisfy the expected demand for power and evaluating each system configuration on the merits of its reliability and cost (1). Solving the optimal mix problem by a simple analysis of the break-even point between the various units in an electric power system (4) can also be included in the category of marginal cost comparison methods.

The capability of performing marginal analysis is an important feature in any decision-making process. It is especially essential in the electric power industry in view of the magnitude and complexity of the investment-decision problems. For any investment alternative, it is also important to perform various types of marginal analyses so as to obtain more insight into the relationships between the various elements of the system, thereby providing useful information for decision making. In particular, marginal analyses are required for the following purposes:

1. sensitivity analysis to determine how sensitive the system is to various changes in the data and the parameters of the system;
2. trade-off analysis to balance conflicting parameters or objectives one against the other;
3. peak analysis to study the behavior of the system in peak hours, an analysis most important for reliability and cost studies as well as rate structure policies;
4. evaluation of bids for power plants to enable decision makers to trade off generating units, which usually come in different sizes, forced outages rates, maintenance requirements, etc.;
5. contingency analysis to study the behavior of the system under contingent and uncertain events, such as sudden increase in load and delay in installing power plants.

Comprehensive models, on the other hand, address problems of a much wider scope, and over a longer period of time. Obviously, these models are not as refined as marginal analysis methods, as various degrees of simplifications and assumptions are required to render the models tractable and solvable. Mathematical programming models, including linear, non-linear, mixed integer, and stochastic, have been used to approach the long-range investment program in the electric power industry. A survey of the types of models used in determining least cost investment in electricity supply appears in (1).

2.3 The Generating System

This book deals exclusively with the generating part of the system. Generating facilities are the largest single item of cost in an electric power system. It should be remembered, however, that electric system operation is also dependent on the proper performance of transmission and distribution equipment.

One should also note the difference between static analysis and analysis that includes spinning reserve. The former is concerned with long-term evaluation of reliability with no attention given to the operating conditions of the system at any specific point in time. The latter is concerned with short-term evaluation (covering perhaps an hour or a day), for which the operating characteristics of the power system at any specific point in time play a major role. The main concern of the present work is with long-term analysis of the generation system. Methods of evaluating spinning reliability can be found in (6). In the following, reliability will always mean static, or long-term, reliability; the word "static" will be omitted for simplicity.

3

Electric Power Generating Systems—Basic Concepts

3.1 Demand and Supply of Electricity

The Demand

The demand for electricity varies according to the time of day, the day of the week, and the season. During nighttime, electricity demand is generally lower than during the day. On very hot or very cold days, when extensive heating or cooling is needed, electricity consumption obviously will increase.

A load–duration curve (LDC) is defined as showing the amount of time that any given overall load level is exceeded. The LDC is one of the most important tools in the analysis of electric power systems. It has been utilized for various purposes, such as estimating the operating cost of a power system, predicting the amount of energy delivered by each unit, and calculating reliability measures.

By its definition the LDC is a function whose abscissa specifies the number of hours, in a given period, usually a year, during which customers' demand for power equals or exceeds the associated demand level on the ordinate. Denoting by P the demand variable and by p a particular value of the demand, $T(p)$ is the number of hours during which $P \geq p$. A typical LDC is exhibited in figure 3.1. This graph will be used later for the various examples. In this graph p is expressed as a percentage of the annual peak demand.

By normalizing the time variable, the value at any point on the abscissa can be thought of as the probability that the correspending load will be equaled or exceeded. By reversing the role of the axes, the LDC can be interpreted as the complementary cumulative density (distribution) function of customers' demand. The cumulative density function (CDF), denoted by $F(p)$, is thus given by

$$F(p) = 1 - t(p), \tag{3.1}$$

where $t(p)$ is the proportion of time during which $P \geq p$, and is equal to the normalized value of $T(p)$ (in the case of a yearly LDC, $t(p) = T(p)/8760$).

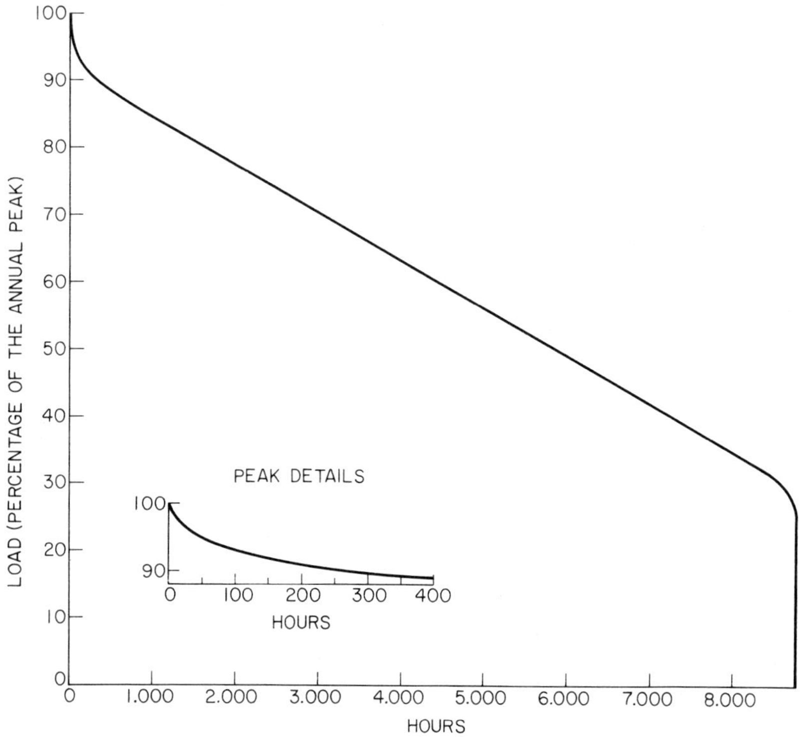

Figure 3.1 Typical load–duration curve (yearly).

The load density function $f(p)$ is obtained from (3.1) by differentiation:

$$f(p) = -\frac{dt(p)}{dp} = -t'(p). \tag{3.2}$$

Figure 3.2 exhibits these curves. Throughout this book we shall use $t(p)$ to denote the LDC. Whenever hours, rather than time fraction, are desired the results should be multiplied by the duration of the analyzed period.

By the manner in which the curve is constructed, the LDC is a monotone decreasing function, implying a one-to-one correspondence between customers' demand p_D and the time $t(p_D)$. An empirical LDC should be considered as one realization of a random function. In practice it is common to consider this realization as an estimate of the expectation of the random function, and sometimes to account for the variation about this expectation by assuming the existence of a random component p_R. The random component represents the forecast error and the consuming-dependent load variations and it is assumed that every value of p_R has a probability density $P(p_R)$. These are demonstrated graphically in figure 3.3.

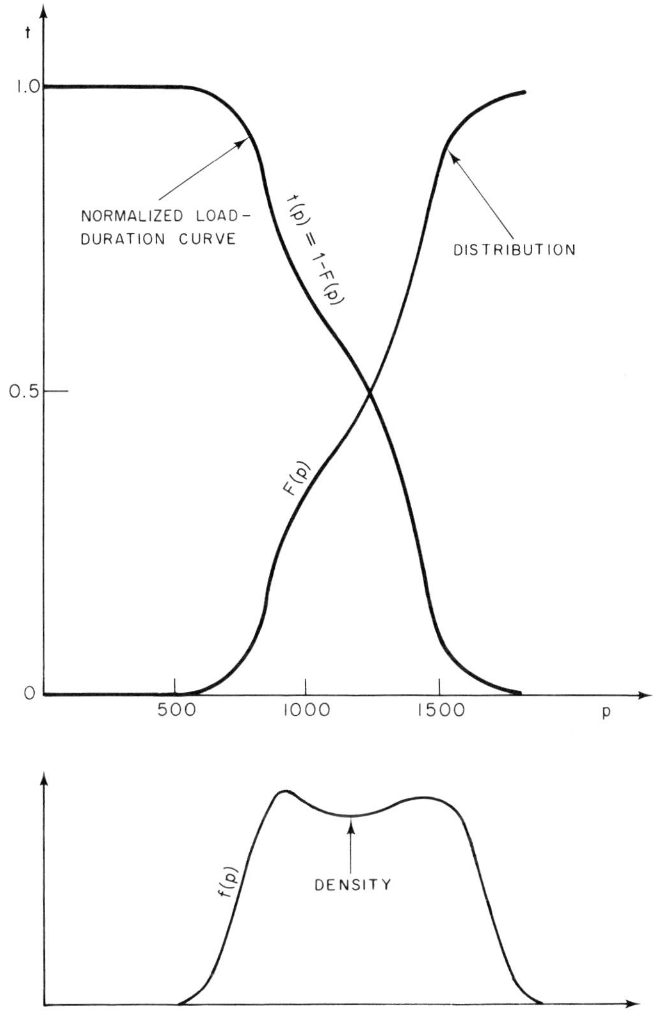

Figure 3.2 The load–duration curve as a cumulative distribution function and the corresponding density function.

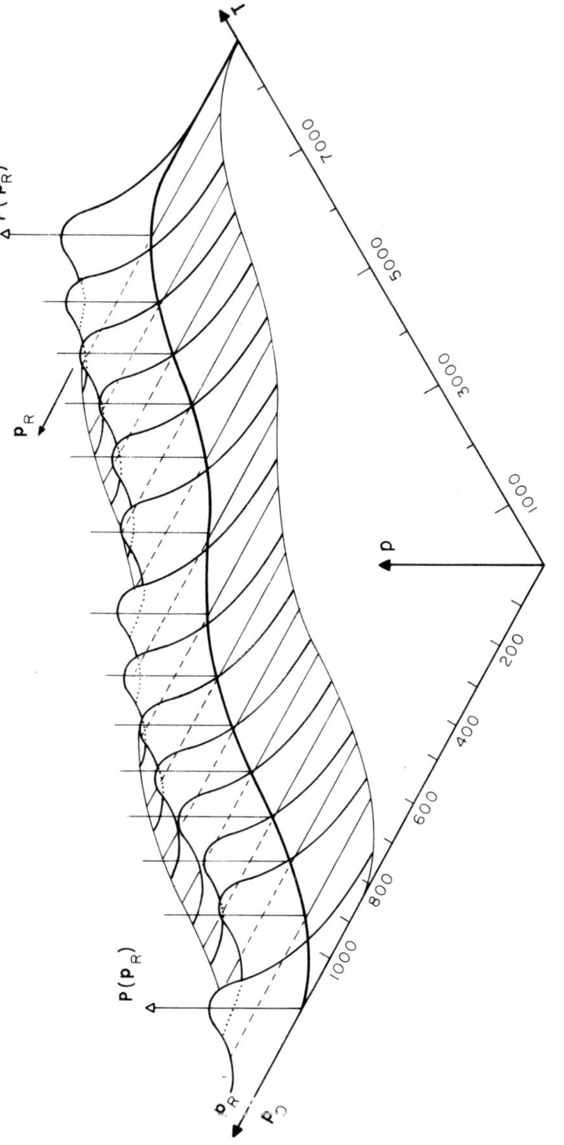

Figure 3.3 Load–duration curve including component of random demand.

Supply and the Probabilistic Nature of Electric Energy Generating Systems

The electricity demand is satisfied by the electric generation system. The system consists of various different generating units, and is characterized by the following parameters:

- number of units;
- size and type of each unit;
- reliability of each unit;
- fuel consumption of each unit; and
- capital cost of each unit.

Whenever the supply does not meet the demand, a shortage is experienced.

Several factors lead to shortages of electric power. These are forced outages of generating units, scheduled preventive maintenance, generator deratings, and deviations of demand from expectations. Forced outages result when a unit goes out of service because of unplanned reasons such as breakdown, malfunction of equipment, etc., and are usually of a random nature. Preventive maintenance, on the other hand, is a planned event, aimed at keeping the generating units in proper running condition. Deratings of generators occur as a consequence of equipment problems and changes in operating conditions, as well as the age of the equipment. And finally, deviations of demand from expectations are due to the random nature of the demand for power, which varies with the time of day, day of the week, and season of the year.

In order to ensure against an excessive amount of shortages and deliver a proper quality of service, electric utilities must maintain an extra generating capacity above the expected peak period demand, called *reserve*. The reserve capacity is actually a safety margin to provide for the day-to-day variations in the available capacity and deviations of demand from estimates. Despite the reserve, there is a small but measurable probability that the available generating capacity will be insufficient on a given day to accommodate the demand for power, resulting in a shortage of electricity. Obviously, the larger the reserve margin, the less the chances of running into shortage situations. The ability of the system to meet the demand for power at any given point in time is referred to as the *reliability of the system*. Most reliability standards are measured in terms of the probability of failing to meet the anticipated load at a usable voltage level and without emergency help from other systems. It is the required reliability standard that actually determines the amount of reserve to be maintained in the system. A utility might therefore ensure a very high reliability standard by maintaining a large reserve margin. However, this is a very expensive proposition, as it involves large investments in extra installed capacity that will actually be used only a small fraction of the time. The greater the reliability designed into the system, the greater will be the cost of meeting this reliability constraint, but, on the other hand, the larger will be the benefits in terms of better service, less interruptions, less inconveniences to customers, etc. Consequently, the question of the amount of reserve to be maintained in any given system depends on the trade-offs between the cost of providing the extra reliability and the benefits accruing from the higher standards. This is one of the most subtle and complex problems in electric power system analysis, as it depends on so many factors, including the generation characteristics of the system, the characteristics of the demand for power, the types of customers served by the system and their sensitivity to power supply interruptions, economic conditions of the country, and political factors.

Due to the vital nature of electricity, both to our economic and personal well being, high standards are usually placed by developed countries on the reliability performance of electricity supply. For example, most U.S. electric utilities are designed on the assumption that there should be no more than 1 day of generation shortages in 10 years. This design target does not imply a full day of shortages once every 10 years. Rather, the total accumulated time of interruptions should not exceed 1 day in 10 years, or, equivalently, approximately $2\frac{1}{2}$ hours per year. This is a very strict design criterion even for developed countries. In fact it has been recently shown that the 1-day-in-10-years reliability design target is economically unjustified, and could reasonably be reduced without an adverse impact on the economy (32). Reliability design targets should in fact be tailored to the particular economy involved and suited to the type of customers served by the system.

3.2 Unit Reliability

Forced Outages Rates

Unit reliability is one of the major factors influencing the reliability of a system, as it determines the frequency with which a unit is likely to be out of service because of an unplanned outage. The lower is the reliability of a given unit, the more frequently it is likely to break down, and the poorer will be the reliability of the system. Unit reliability itself is affected by many factors which will be discussed in the next section.

Unlike maintenance requirements, forced outages are random variables, and their occurrence and duration are unpredictable. The failure pattern of a given generating unit is usually described by its *forced outages rate* (FOR), computed as the percentage of the unit's down time relative to the total service plus down time:

$$\text{FOR} = \frac{\text{forced outages hours}}{\text{forced outages hours} + \text{service hours}} \times 100, \tag{3.3}$$

with the forced outages hours defined as the time in hours during which a unit is unavailable due to a forced outage, and the service hours being the total number of hours the unit is actually operated with breakers closed to the station bus. Forced outages are defined, for that matter, as the occurrence of a component failure or other condition which requires that the unit be removed from service immediately, or up to and including the very next weekend. The FOR is an estimate of the conditional outage probability of the unit, that is, the estimated probability of a forced outage given that the unit is needed for generation. The unconditional, or overall, outage probability is obtained by multiplying the conditional probability by the fraction of time during which the unit is needed for generation (multiplying the FOR by the fraction of time the unit is either down or in service relative to the total duration of the time involved, exclusive of maintenance time). Whenever outage probability, denoted by q, is mentioned in this book it is to be interpreted as the conditional outage probability interchangeable with the FOR. The unconditional outage probability will be denoted by q^*.

Unit Availability

Forced outages are one factor that might reduce the available capacity at any given time, the other being maintenance requirements. Obviously, the larger the number of units out of service at a particular point in time for maintenance purposes, the smaller the available installed capacity, and the larger the chance of shortages. It is customary to express maintenance and forced outages requirements of a given unit by means of a combined index, the *operating availability*, defined by

$$\text{operating availability} = \frac{\text{available hours}}{\text{period hours}} \times 100, \tag{3.4}$$

where the denominator denotes the number of hours in the period involved (say a year) and the numerator denotes the number of hours in the period during which the unit is available for service, whether or not it is actually in service. This includes both service hours with breakers to the station bus closed and reserve shutdown hours during which the unit is shut down but is available for service.

A related measure is the *unit availability* factor, defined by

$$\text{unit availability factor} = \frac{\text{on-line time (hours)}}{\text{duration of period (hours)}} \times 100, \tag{3.5}$$

where the denominator is as defined above and the numerator denotes only the time (in hours) during which the generator was operating with breakers closed to the station bus.

We note that the latter measure is also affected by the position of the unit in the merit order. The higher the unit is in the merit order, the less it will be operated, and the smaller will be the on-line time and thus the unit availability factor. In fact, it is this factor that should be considered in evaluating the impact of the unit on system reliability. Clearly, the smaller the unit's availability factor, the less the impact of the unit on the reliability performance of the system. It follows, then, that base-load units bear larger impact on system reliability than peaking units because their availability factor is higher.

Factors Affecting Unit Reliability

Unit Type Each type of unit has its typical FOR. Hydro units usually have the smallest FOR, then fossil-fueled, nuclear, pumped-storage, and gas turbines, in increasing order. However, since FOR also varies according to size, we may find small nuclear units with FOR lower than large fossil units.

Unit Size The larger the unit size, the poorer the unit reliability. The increase in FOR is usually attributed to the increased sophistication, complex elements, and higher degree of automation associated with the larger units.

Unit Manufacturer FOR varies among manufacturers even for units of the same type and size, due to differing technologies and production processes. Data about the predicted FOR of a unit are usually provided by manufacturers and could be used for assessing the resulting system reliability.

Unit Age FOR also varies with the age of the equipment. In the first 2 or 3 years after installation, the FOR of the unit will tend to improve (i.e., get smaller) because of a learning and debugging process. Then the FOR will remain at a fairly constant level, up to a period when it starts to increase due to aging of the equipment.

Design Vintage Newly designed units are usually associated with larger FOR. As more experience is obtained over the years by both manufacturers and users, corrective measures can be taken to eliminate equipment and operating problems, resulting in improved FOR.

Maintenance Planned maintenance aimed at keeping generating units at proper running conditions and routine inspection to detect possible malfunctions result in higher unit reliabilitv. In fact, scheduled maintenance is indeed one of the cheapest and most effective ways to decrease the FOR of a unit.

Parts and Replacements Stocks Maintaining an adequate inventory of replacements and spare parts is essential for bringing units back to service in the shortest possible time. Shorter repair time implies smaller FOR. Standardization of units might be helpful in reducing outage costs as it requires stocking a smaller quantity of spare parts. Another effective approach, in the case of high-cost replacements having low-need probability, is the creation of a pool of utilities sharing a common stock of replacements.

Regulatory and Environmental Factors Installation of various equipment, such as scrubbers and precipitators, to meet environmental standards, results in increasing the FOR of the unit. Various other regulations, which require that units be shut down at certain times to comply with air pollution standards, also imply a lower reliability for the units. The impact of environmental regulations on system reliability is particularly high since it affects large fossil-fueled plants, which are usually the most polluting units.

Design and Engineering Factors Design and engineering considerations might affect the unit reliability. Units designed in a manner that enables quick detection and easy location of malfunction are associated with smaller FOR.

Procurement and Construction Phase Faulty manufacture or improper installation will result in equipment failure, and therefore higher FOR. Hence proper choice of equipment and components during the procurement phase, and careful inspection during the construction phase, might guarantee better unit reliability.

Quality of Operation Human errors are responsible for many forced outages. Proper planning of operations, comprehensive training programs, and careful control will improve the FOR of the unit.

3.3 The Random Nature of the Level of Available Capacity

Given the outage probabilities and assuming that (1) each generating unit can be in either one of two possible states—operative and thus capable of full power generation, or nonoperative and incapable of delivering any energy—and (2) the failure pattern of a given unit is independent of its generation level, the deterministic and random components of customers'

Table 3.1 Outage Probabilities for the Three-Unit Example

Case	Unit 400 MW	300 MW	250 MW	Available capacity (MW)	Probability	Capacity in outage (MW)
0	Up	Up	Up	950	$0.94 \times 0.96 \times 0.98 = 0.8844$	0
1	Up	Up	Down	700	$0.94 \times 0.96 \times 0.02 = 0.0180$	250
2	Up	Down	Up	650	$0.94 \times 0.04 \times 0.98 = 0.0368$	300
3	Down	Up	Up	550	$0.06 \times 0.96 \times 0.98 = 0.0564$	400
4	Up	Down	Down	400	$0.94 \times 0.04 \times 0.02 = 0.0008$	550
5	Down	Up	Down	300	$0.06 \times 0.96 \times 0.02 = 0.0012$	650
6	Down	Down	Up	250	$0.06 \times 0.04 \times 0.98 = 0.0024$	700
7	Down	Down	Down	0	$0.06 \times 0.04 \times 0.02 = 0.00005$	950

load, and the outage pattern of other units in the system, the generation capacity that is out of service due to forced outages is multinomially distributed with the outage probabilities as parameters. The distributed level of available capacity takes on 2^N values, equal to the number of combinations of up and down (due to forced outages) units in an N-unit system.

In the following example, a three-unit system will be assumed with capacities 250 MW, 300 MW, and 400 MW, and outage probabilities 0.06, 0.04, and 0.02, respectively. Eight combinations of up and down units are possible in this case. Table 3.1 demonstrates the calculation process of the available capacities probabilities. This example will be used throughout this book. The reader should be aware, however, that for the purpose of clarity we have selected a "synthetic," oversimplified example.

3.4 Loading Procedures

Incremental (or Marginal) Cost Loading and Merit Order Loading

Given the demand for power and the units available, a loading procedure is applied. The loading procedure is the manner by which the various units are assigned to generation in order to meet the instantaneous demand for electric power while minimizing operating costs. As such, the loading procedure determines the number, type, and production level of the units which are loaded to generation at any given time. It is therefore a most important factor in calculating the operating cost of a given unit. The major operating cost component, excluding start-up, shutdown, and transmission losses, is fuel cost. Fuel consumption is a nonlinear function of the unit's output; in particular, it has been empirically demonstrated (12) that the cost function of generating units in a power system, with the possible exception of hydroelectric and pumped-storage units, is an increasing function, exhibiting an increasing marginal cost. Under these conditions, it can be proven (12) that the loading procedure which yields the minimum operating cost for the system is the one which equates the marginal costs of the units in the system for any demand level. This loading procedure, widely implemented by electric power companies, is known as the *incremental load* (sometimes the *economic dispatch*) procedure. The incremental loading method will be reviewed more extensively in chapter 5. It is difficult, however, to calculate in advance the fuel cost in a system loaded according to the incremental loading procedure as it depends on the dynamic changes in demand and availability of units, which can be taken into account only in an hour-by-hour analysis. Indeed, simulation models which are capable of an hourly analysis have thus far provided the only means for predicting the operating cost of the system based on incremental loading.

An alternative assumption regarding the loading procedure, which is common in analyzing operating cost of systems since it avoids the difficulties presented by incremental loading, is *merit order* loading. According to this method, units are assumed to be loaded for generation in order of their average production costs, which serve as an indicator of a unit's merit order. The most efficient unit, i.e., the unit with the lowest average production cost, is loaded first and operated at its rated capacity; the next most efficient unit is then loaded to generation at its rated capacity, and so on until demand is satisfied. This procedure is certainly much more convenient for cost calculation; and in conjunction with the LDC it yields a very efficient algorithm for predicting the operating cost of the system. The procedure, however, yields only approximate results.

Booth (9) has suggested an improvement of the analysis according to merit order loading by dividing the total capacity of a given unit into two capacity blocks, each with a different average production cost, which are then placed in nonadjacent positions in the merit order. This loading procedure results in greater accuracy in calculating the operating cost of the system. Obviously, the larger the number of blocks for a given unit, the more accurate the operating cost predictions. In fact, incremental loading is basically a further refinement of the two-block method, with the capacity of a unit divided into infinitely many infinitesimal blocks; the cost of each is the incremental cost of producing the energy in that block.

The Effect of Forced Outages on the Loading of Units

In merit order loading, an additional increment in demand at any hour will be picked up by the last unit in the merit order which is scheduled for operation at that hour, or, if the last unit is loaded at full capacity, by the next available unit in the merit order. For example, in figure 3.4 we describe the merit order loading for a power system, with unit 1 denoting the most efficient unit, unit 2 the second most efficient unit, and so on. The hour t in figure 3.5A will therefore be served by unit 3 as the last unit.

In electric power systems with no forced outages, it is very easy to calculate the operating cost. However, in the case when the generating units are assumed to be subject to forced outages, different combinations of units might be assigned to generation for the very same load, thereby giving rise to different operating costs. Going back to figure 3.4, suppose that unit 1 is down because of a forced outage. Then, all the other units will occupy a lower position in the merit order, as shown in figure 3.5B with unit 5 now serving as the last supplying unit for hour t. Hence the operating cost for hour t, for this realization, will be different. Assuming the standard two-state model for each generating unit due to forced outages (operative and thus capable of full power delivery, or nonoperative and incapable of producing any electricity), as many as 2^N realizations are possible, equal to the number of combinations of up and down units in an N-unit system, each resulting in a different operating cost.

An effort to calculate all these 2^N variations will result in an excessive amount of calculation, which increases exponentially with the number of units in the system.

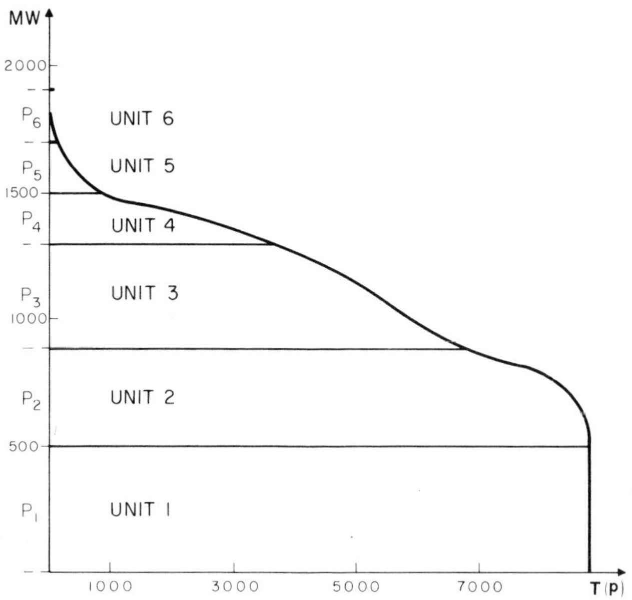

Figure 3.4 Loading according to the merit order approach.

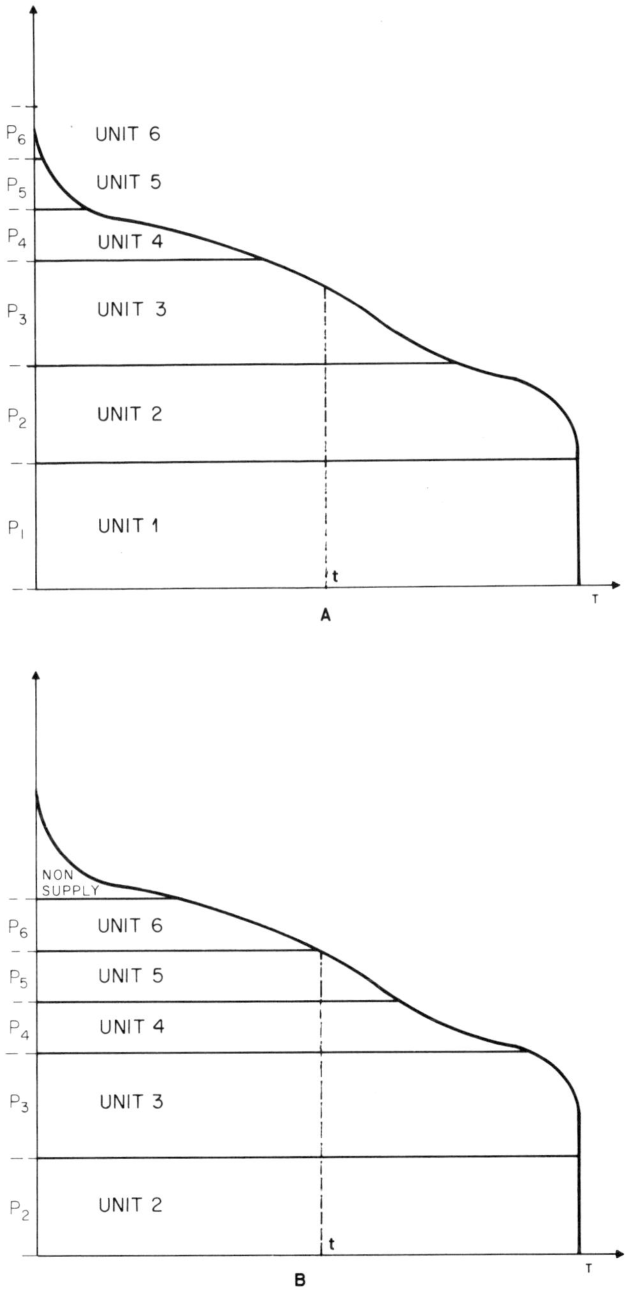

Figure 3.5 Two possible states of a system loaded according to merit order. (A) Unit 1 in; (B) unit 1 out.

The Effective Capacity Loading Method and Its Deficiencies

An approximate procedure to account for power supply uncertainty in calculating the cost of meeting an increment of energy demand is to use the effective capacity approach (28, 29) in conjunction with merit order loading. According to this approach, forced outages and maintenance requirements are taken into account by reducing the rated capacity of each unit by a certain factor, reflecting the average fraction of time during which the generating unit is not available for production (figure 3.6). In this case the costs associated with the operation equal the average production cost, with the units loaded to generation in the merit order, at their effective rather than their rated capacity (23). While the approach is indeed very simple, it does not account for maintenance and forced outages requirements in a realistic manner since it assumes an even distribution of outages over the year, an assumption which does not hold true in reality. In fact, it has been verified that the effective capacity approach tends to underestimate the production expenses of an electric power utility (28, 29). Prices based on marginal energy costs calculated using this approach are likely to be underestimates, and would not recover the costs incurred in producing electricity.

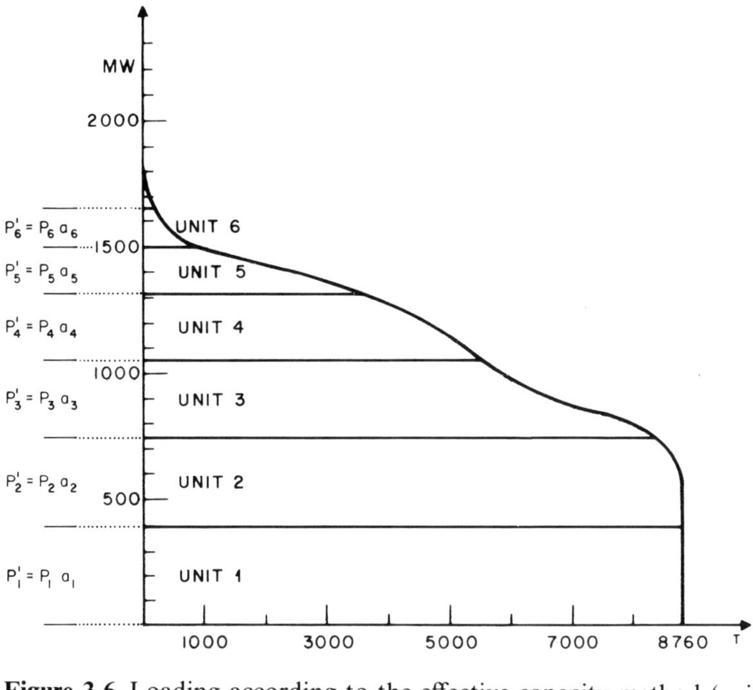

Figure 3.6 Loading according to the effective capacity method (a_i is the unit's effective capacity factor and P'_i is the unit's effective capacity).

4

Reliability Analysis and Reserve Planning in Electric Systems

4.1 Introduction

Due to the nature of electricity and the dependence of our society on its uninterrupted supply, reliability is one of the most important design criteria of electric power systems. System reliability depends upon many factors, including the amount of installed capacity, unit size, unit reliability, maintenance requirements, interconnections, dispatch facilities, load forecasting errors, and shape of load. Electric power systems maintain reserve generating capacity as a safety margin against the day-to-day fluctuations of the operative installed capacity and deviations of customers' demand. The higher the reserve margin, the higher the reliability performance of the system. However, providing excess electric generation capacity for reliability purposes has an economic cost, in fact a substantial one. Large investments are required for installing extra capacity that is utilized only a small fraction of the time. Also, the higher the reliability designed into the systems components, the higher the associated costs. On the other hand, higher reliability provides some benefits by reducing the cost of nonsupply and adding to the value of the electric service. The selection of the optimal reliability design target should bear an appropriate relationship to the benefits accruing to the society by higher reliability standards versus the additional costs of providing these standards. However, because of the difficulties in estimating the social benefits of reliability, other methods, such as contingency and probabilistic methods, are commonly used for determining the level of reserve to be maintained by the system. In determining these requirements, it is also necessary to consider the potentialities inherent in load-management schemes and interconnections as possible means of improving reliability performance at a substantial saving in costs.

4.2 The Reserve Planning Problem

The reserve planning problem is concerned with determining the optimal level of reserve margin to be maintained by the system. The optimal reserve depends on various generation and demand characteristics of the system, some of which have been described in the previous chapter, as well as the cost–benefit ratio of providing extra reliability. Furthermore, adding extra reserve is not the only means of meeting reserve requirements and satisfying reliability constraints. Other methods, which might be cheaper and more efficient than adding extra reserve, call for changing the generation characteristics of the system and/or affecting the demand for power by shaving peak demand and improving the load factor of the system, both of which result in better reliability at a reduced cost. It is therefore necessary to look upon the reserve planning problem from a more general point of view by giving consideration to all factors affecting system reliability and accounting for the interaction among these factors and their impact on the system costs.

4.3 Parameters Affecting System Performance and Their Impact on Costs

Installed Capacity

Adding extra capacity is a common way of increasing system reliability. Figure 4.1 shows the relationship between installed capacity and system reliability. If an electric power company is to provide for higher reliability standards by increasing the reserve margin, it should have to increase investment in adding extra reserve that will actually deliver a very small amount of energy. Or, equivalently, reducing reliability standards (increasing shortages) would imply major savings in investment.For example, in a reliability study of the electric systems in New York State it was found that "if the New York system were to shift from a one day in 10 years criterion to a one day in one year, there would be a saving of capital amounting to \$20 million in the first year and rising to \$36 million seven years later, then dropping to \$12 million five years later" (21), a substantial savings. In view of the increasing construction costs of power plants, extra capacity as a way of meeting reliability standards is becoming less economical.

Composition of Sizes of Units in System

Usually the smaller the units in the system, the more reliable the system for the same aggregate capacity. Reliability considerations alone will therefore drive the system toward smaller units. Considerations of economies of scale, on the other hand, will tend to drive the system toward larger units. Trade-off analysis is therefore required in order to balance the two factors one against the other and determine the optimal unit size. The trade-off is between choosing smaller, and more expensive, units with lower reserve, or reducing the number of units and installing extra reserve. This trade-off

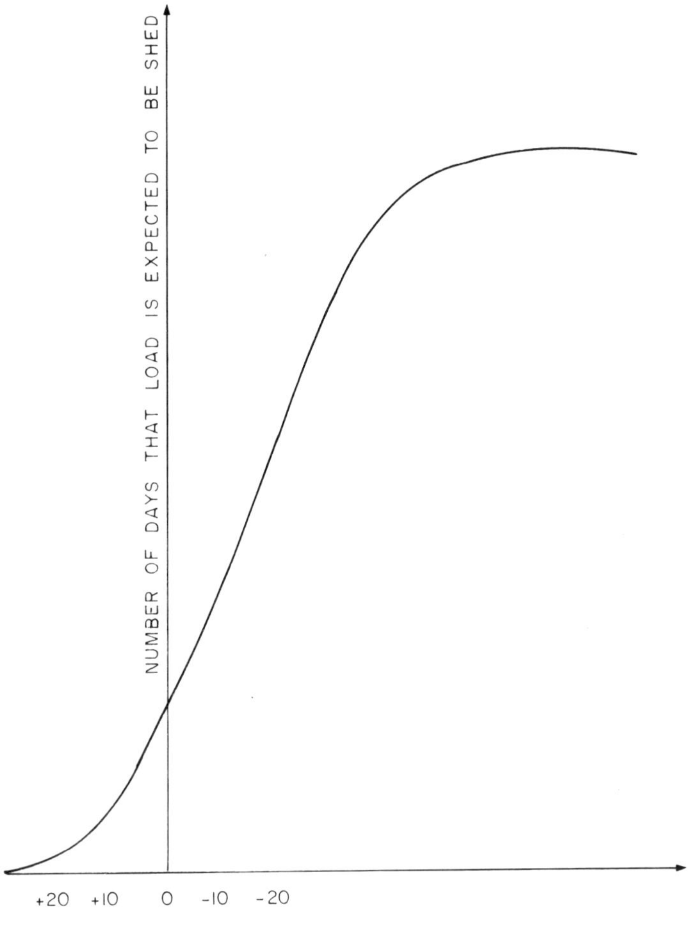

Figure 4.1 Number of days that load is expected to be shed versus generation reserve.

analysis is especially important in small systems with only a few units. These systems are more sensitive to changes in unit size than large systems. Of special consideration are nuclear units, which, because of economy of scale, are not available in sizes less than 600 MW. Hence installing a large nuclear unit might immediately call for installing extra reserve.

Some rule-of-thumb methods, based on experience and technical intuition, have been suggested for determining the size of the new unit. For example, no single unit should exceed 7% of the system's peak demand.

While rule-of-thumb methods are appropriate as general guidelines, an overall optimization is required in order to find the exact optimal unit size for any specific system.

Generator Characteristics

System reliability is also affected by the generation characteristics of the system. A method of meeting reliability standards, possibly cheaper than adding extra reserve, is to select the appropriate characteristics of the units in the system, such as their size, FOR, and maintenance schedule. Since it is difficult to change the size and FOR of existing units, this alternative should be considered particularly when purchasing new units and evaluating bids for power plants.

Interconnection

Interconnection is a common way to increase system reliability and to save in investment costs. Since occurrences of outages in any two systems are not necessarily dependent, when one system suffers from capacity deficiencies it might be able to purchase power from neighboring systems which happen to possess excess supply over demand at that time. Special advantages exist if the load shapes of the systems are different and especially if there is a time lag between the peak demand hours of the systems. Consequently, reserve requirements of an electric power system interconnected to other systems are smaller than if the system were not able to buy power or sell it to its neighbors. Whenever possible, interconnection to other systems provides a major cost-effective means of meeting reliability standards, at a substantial savings in cost, by making it possible for the power companies involved to share their reserve margin. Besides increasing system reliability and reducing requirements for reserve margin, interconnection also provides for installation of larger units without a need to back them up with additional peaking facilities, thus further reducing investment and operating costs. If there is a time phase between the peak demand periods of the interconnected systems, the overall load factor of the systems will be improved, with major cost savings to all parties. The economics of interconnection should be evaluated on the basis of its contribution to system reliability and installation and operating costs as compared to the connecting tie-lines costs, requiring again a trade-off analysis between the interests involved.

Maintenance Scheduling

Preventive maintenance is required for keeping generating units operating reliably and at top efficiency. Proper maintenance scheduling is of importance in maintaining system reliability. On the other hand, maintenance activity removes capacity from the system, thus exposing the system to an increased risk while the maintenance activity lasts. Unlike forced outages, maintenance operations can be planned in advance and scheduled so as to minimize the chances of shortages. Two criteria are usually utilized for scheduling maintenance operations over the year: (1) leveling reserve capacity throughout the year; and (2) leveling risk of losing load. In the first approach maintenance is scheduled so as to level the reserve capacity by placing maintenance operations into periods of slack demand. In the second approach, we attempt to level the shortage probability at every maintenance period, with a maintenance period defined as a period during which a certain combination of units is undergoing maintenance. The two approaches do not necessarily result in the same system reliability. The second approach usually provides a better reliability for the system because it accounts for the special characteristics of the generating units as well as the demand for power at each maintenance period. However, it is much more difficult to execute.

In figure 4.2 we show the actual durations of various capacities under maintenance of a certain system. It is interesting to note that in spite of the variations, the general shape of the curve is similar. In figure 4.3 we present a typical maintenance schedule. The large units (1 and 9) are scheduled to be maintained during low demand, smaller units during higher demand.

Load Forecasting Quality

Due to the lead time involved in the installation of new generating units, which might last from a few months for gas turbines to more than a decade for nuclear units, investment decisions must be made several years before the commissioning of the units. A major factor in this decision-making process is load forecasting, both of the shape and the peak of the demand for electric power. The quality of load forecasting bears considerably on the reliability performance of a system, with poor forecasting resulting in either unnecessary investments in reserve margins, or in less than desired system reliability.

4.4 Reserve Planning Methods

Three groups of methods are available for reserve planning: (1) nonprobabilistic (contingency) methods, which set the reserve level at some percentage of peak demand or generating capacity; (2) probabilistic methods (6), which seek to find the reserve level that yields a given level of reliability (mostly the loss-of-load probability—LOLP); and (3) economic methods (19), which seek to evaluate the most economical level of reliability to maintain in the system when determining the reserve requirement.

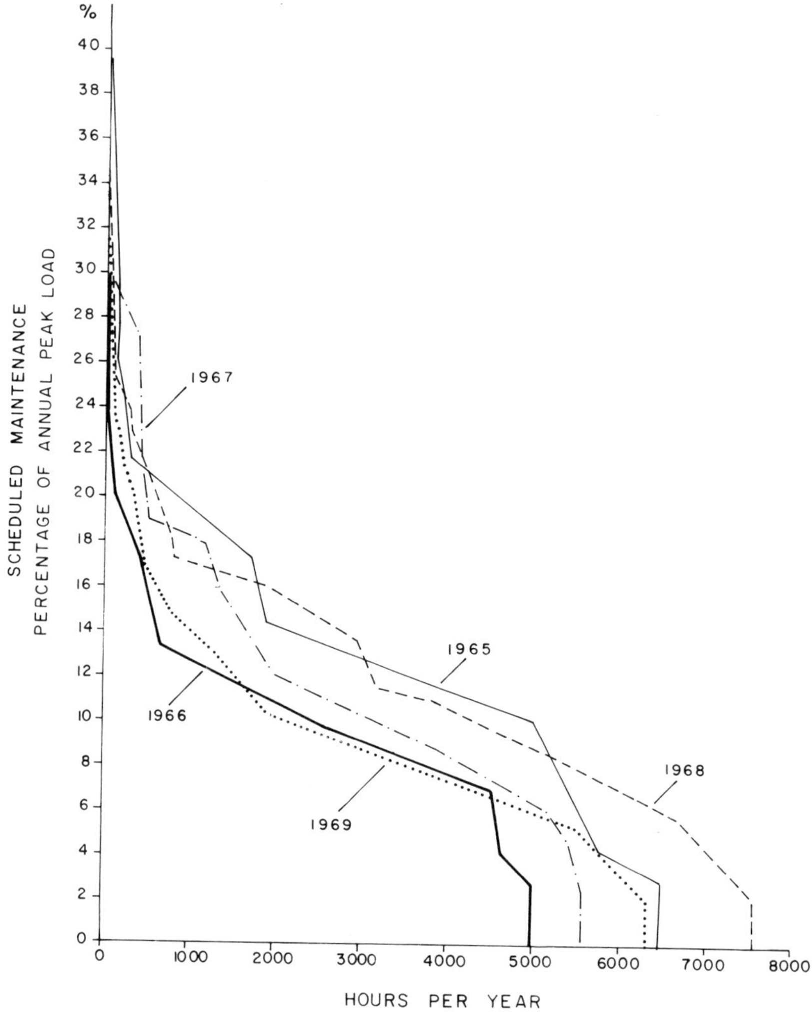

Figure 4.2 Duration of capacity under scheduled maintenance of a specific system.

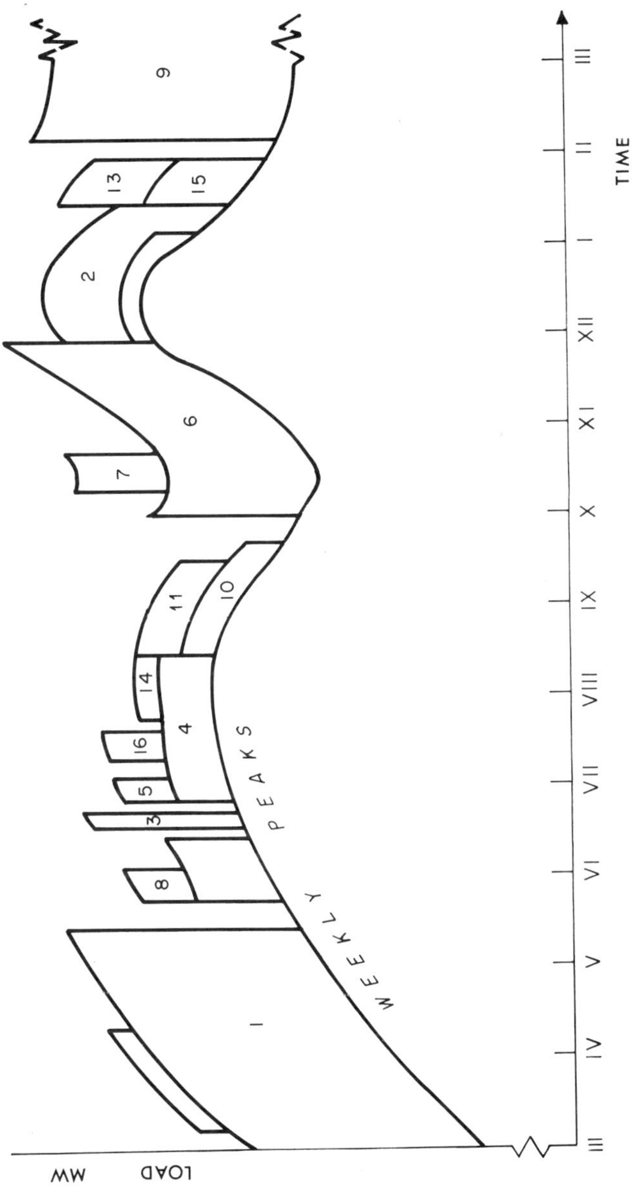

Figure 4.3 Typical maintenance schedule in relation to weekly demand peaks.

Contingency Methods

Under these methods reserve margin is chosen to satisfy a certain contingency test, regardless of the reliability design target. The most common methods are

1. *The standard percentage reserve.* Under this method, reserve level is determined at some fixed percentage of the forecasted peak load of the system. The exact percentage varies in the range 15–25%, and is determined by the utility's own experience or by adopting other utilities' standards.
2. *The largest unit.* In order to allow for power supply even if the largest unit is out of service, reserve is set equal to the capacity of the largest unit. This method might be somewhat better than the previous one, since reserve is increased if a larger unit is introduced, but still this criterion does not guarantee that the reliability target is met.
3. *The largest unit plus forecast error.* This method takes into account possible errors in forecasting the demand for power by increasing the reserve margin by some percentage of the peak demand. A common forecast error allowance is 5%. This level of reserve will indeed allow uninterrupted supply even if the largest unit is out and demand exceeds previous predictions, by a certain percentage, but it still suffers from all the deficiencies mentioned earlier.

When using contigency methods, system composition (number, size and FOR of units) as well as the demand characteristics are not reflected in the required reserve, in spite of their impact on system reliability. Use of these methods will usually result in either excess generating capacity and excess capital costs, or inadequate reliability due to insufficient installed capacity.

Probabilistic Methods

These methods take into account, in a probabilistic manner, the random nature of both the demand and the supply of power, and therefore yield better results. Typically a reliability design target is set, such as LOLP of 1 day in 10 years, and then the reserve margin is adjusted to meet the required target. The common methods are

1. *Searching methods using analytical techniques* (6). For any level of reserve margin, the reliability measure of the system is calculated using analytical techniques. If it matches the reliability design target, the process is terminated. Otherwise, a new reserve margin is chosen, and the process is repeated again. Usually, only demand pattern, as represented by the LDC, and forced outages requirements are accounted for by these methods.

2. *Simulation* (14). With simulation, reserve margin is again determined by a trial-and-error procedure; however, instead of using analytical techniques, simulation models are used to evaluate the reliability performance of the system for any reserve margin level. Uncertainties in both demand and supply can easily be incorporated into the simulation process.
3. *The equivalent load–duration curve* (*ELDC*). The model offered in the second part of this book constitutes a substitute for simulation in that it enables the system planner to account for the fluctuations of the installed capacity due to maintenance and forced outages as well as the random deviations of customers' load in an analytical manner. If reserve requirements are provided by peaking facilities (e.g., gas turbines) that work only a small fraction of the time, the reserve margin required, for any level of LOLP, is easily read directly off the ELDC. Otherwise, a trial-and-error procedure is required to find the exact level of reserve.

Economic Methods

Neither the contingency methods nor the probabilistic methods are concerned with the optimal level of LOLP to be pursued by the system. From an economic point of view, the LOLP level (or any other reliability criterion), and therefore the resulting reserve margin, should depend on the cost of providing extra reliability by changing one or more of the system's parameters, on the one hand, versus the benefits accruing to society by the additional reliability, on the other hand. The benefits that accrue because of the provision of extra generating reserve are the savings of nonsupply costs due to reduction in the occurrence of nonsupply situations. Ideally, reserve capacity should be added up to the point at which the marginal expected cost to consumers of the last kilowatt-hour not supplied equals the expected cost of supplying this kilowatt-hour. The cost of supplying the marginal kilowatt-hour, *assuming the last machine does not experience forced outages, and loading to generation is according to the merit order*, is equal to the annual fixed charges of the last machine in the merit order divided by the number of hours it will run at maximum capacity plus its fuel charges. But the number of hours the last machine will run at maximum capacity is equivalent to the system's LOLP. Hence, given the annual fixed charges, the fuel charges, and the cost of unsupplied energy, the optimal level of LOLP can easily be derived. The reserve margin could then be adjusted to meet the optimal level of LOLP.

Two major difficulties arise in this connection. First, as explained previously, it is difficult to forecast the behavior of peak units when using conventional methods, due to the great number of combinations of in and out units and their effect on the use of peak units. In the second part of the book, a method for describing a system's peak in an accurate manner, taking care of this difficulty, is presented. The second major difficulty concerns estimating the cost of nonsupply. First, it involves a quantification of intangible items such as inconvenience to consumers. Also, various consumers might react differently to nonsupply: some will suffer more, some less. In addition, the nonsupply cost is a function of the frequency, severity, and length of loss of load. Notwithstanding these limitations, a few attempts have been made to assess the social costs of unserved energy (10, 21, 27), but the results obtained are still coarse and further studies are required to achieve more accurate estimation. In the absence of reliable estimates of the cost of nonsupply, LOLP levels are usually determined rather arbitrarily, either by experience or convention, leading to reserve margins that might be far from economic equilibrium.

4.5 Reliability Measures

The Loss-of-Load Probability Measure

Several measures have been devised to evaluate the reliability performance of a given electric power system. The simplest and most common of all is the *loss-of-load probability* (LOLP), a quantity which describes the expected accumulated amount of time in a given period, usually a year, during which the system will experience a shortage of energy of more than 0 kW h. The LOLP is usually expressed in terms of days per year, hours per day, or as a percentage of time. When expressed as a fraction of time, the LOLP can be thought of as the probability that there will be a shortage of power of any magnitude in the given period. Hence the name assigned to this measure.

Let us use the system as described in the example of section 3.3. Outage probabilities were calculated there. The load–duration curve of the system is assumed to be the one given in figure 3.1, and 100% of the demand is 800 MW. For every level of available capacity the duration during which demand will exceed this capacity is checked. This duration is the shortage duration for that event, as demonstrated in figure 4.4 (see table 4.1).

Other Measures and Methods

The major disadvantage of the LOLP is that it only indicates the cumulative expected duration of interruptions; yet there might be great differences between the amounts of unserved energy of systems with equal levels of LOLP. A supplementary measure of reliability is the expected amount of unserved energy, referred to as the *loss-of-energy probability* (LOEP). The LOEP indicates the expected total amount of energy required

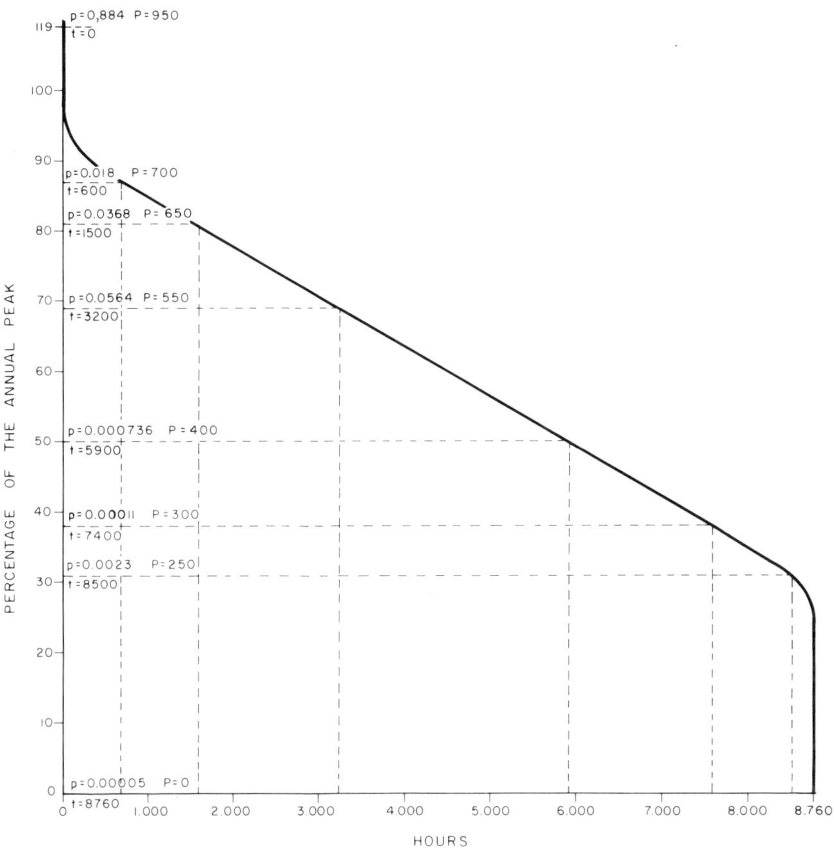

Figure 4.4 Calculating the LOLP. Key: *p*, available capacity; *P*, corresponding probability; *t*, corresponding duration.

by customers and not supplied because of the inability of the system to meet the demand for electric power.

Although indicating the amount of energy which is not supplied, the LOEP does not give any information about the magnitude of demand (kW) that is not satisfied. For example, the same amount of unsupplied energy can be due to extended but small demand interruptions, or shorter but larger demand interruptions. A more sophisticated measure, which provides information on both the magnitude and expected duration of unsatisfied demand, is the distribution of the unsupplied energy, referred to as the *nonsupply distribution*. By definition, the base of the nonsupply distribution equals the LOLP of the system, with the area under the curve

Table 4.1 Calculating the LOLP

Available capacity (MW)	Percentage of peak demand	Probability P_i	Nonsupply duration $t(p_i)$[a]	Contribution to LOLP $t(p_i)P_i$
950	119	0.884	0	0
700	87	0.018	600	10.2
650	81	0.0368	1500	49.2
550	69	0.0564	3200	180.4
400	50	0.0008	5900	4.3
300	38	0.00012	7400	8.1
250	31	0.0024	8500	19.5
0	0	0.00005	8760	0.4
				LOLP $= \sum t(p_i)Pi = 272.1$

a. From figure 1.1.

equal to the LOEP of the system. The relationships between these three reliability measures are described graphically in figure 4.5. All three systems have the same LOLP and systems 2 and 3 have the same LOEP. However, these are not enough to describe the system's reliability. We see that the nonsupply distributions still differ from each other.

An extended version of the latter measure is the *capacity margin distribution*, which relates the difference between the available capacity and the demand for electric power (the capacity margin) to its frequency of occurrence. By definition, the tail of the capacity margin distribution, which corresponds to negative capacity margins, is the nonsupply distribution. The capacity margin distribution is becoming an increasingly important measure of reliability, especially in connection with reliability studies of electric power pools (34). A typical distribution is exhibited in figure 4.6.

Other approaches to measuring reliability which are concerned not only with cumulative duration of nonsupply but also with the frequencies, magnitudes, and duration of nonsupply are the frequency and duration methods (26). These methods are mostly applicable to short-term reliability evaluation, and will not be considered in this book. A comprehensive bibliography on the application of probability methods to evaluating reliability of electric power system performance can be found in (7).

Simulation models provide one common and powerful tool for estimating a system's reliability measures (14). A model of the system is run on a computer, with the various forced outages generated by Monte Carlo methods according to assumed or known outage probabilities. For each time element, the available capacity is compared to the load and the excess

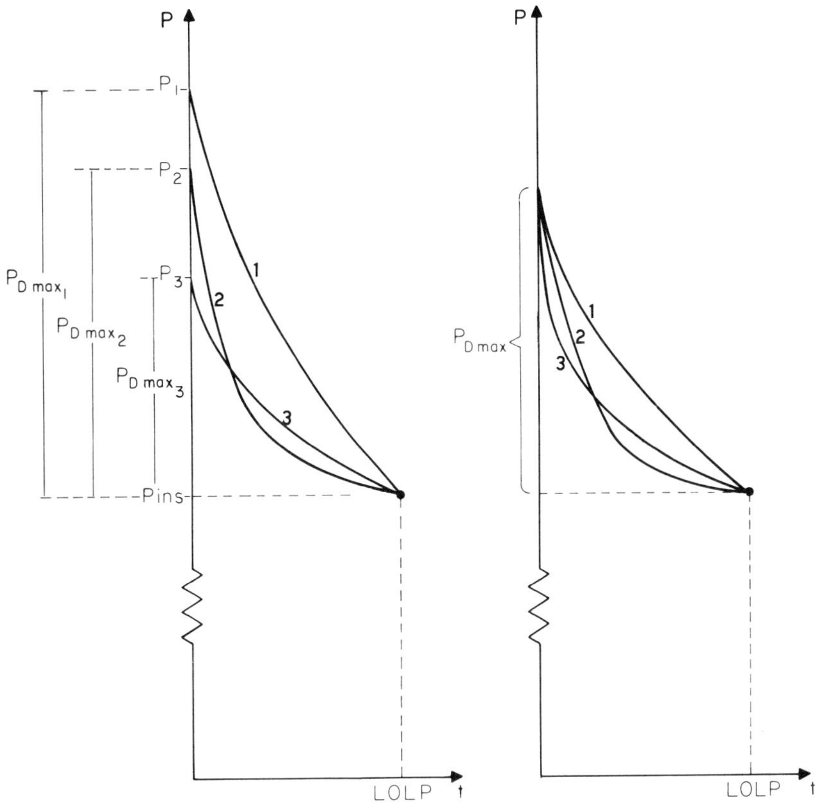

Figure 4.5 Nonsupply distributions. All distributions have same LOLP; cases 2 and 3 have same LOEP. Key: P_i, peak load of system i; $P_{\mathrm{D\,max}\,i}$, the largest magnitude of unserviced demand in system i; P_{ins}, installed capacity.

or shortage of capacity is found. The time periods associated with each excess or shortage level are then registered to yield the required reliability measures.

Various analytical methods, based on the classical load–duration curve, have been devised so far to calculate the LOLP, the LOEP, the nonsupply, and the capacity margin distributions, with forced outages and maintenance requirements incorporated into the curve in an approximate or accurate manner (6). In the second part of the book we develop the approach based on the equivalent load–duration curve, which will also account for random deviations of consumers' demand in calculating these reliability measures.

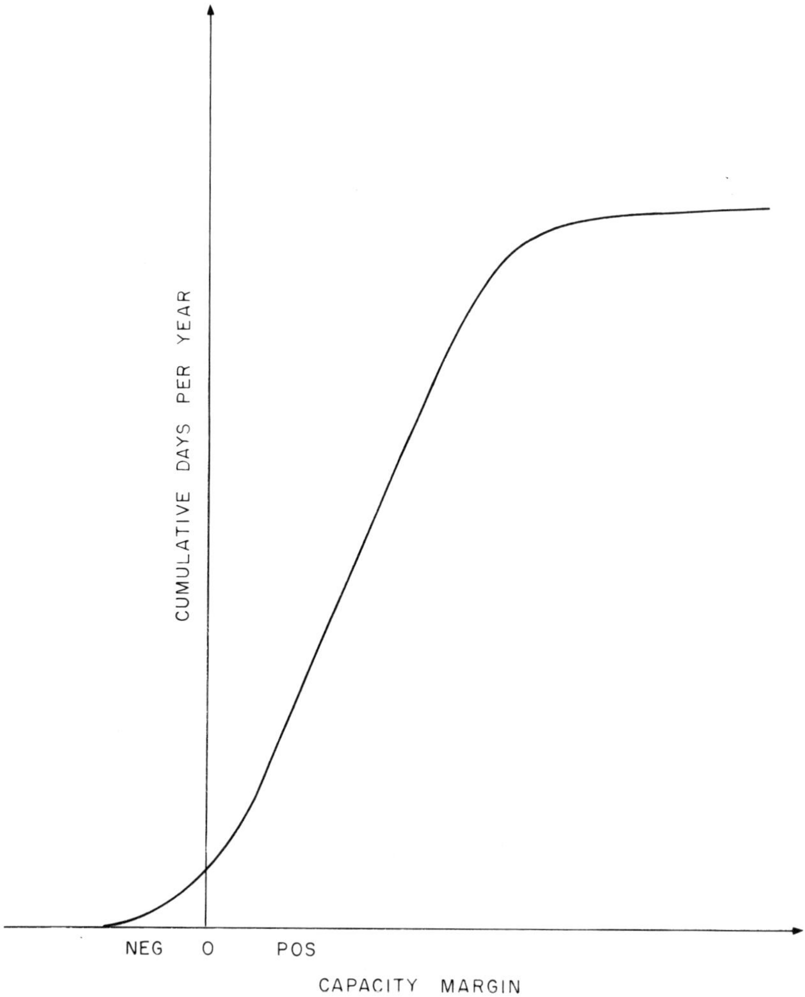

Figure 4.6 Cumulative distribution of daily capacity margin for a typical year.

4.6 Adjusting the Parameters of the System to Maintain a Given Level of Reliability through Load Management

As already mentioned, an alternative way to meet a given reliability level, possibly even cheaper than adding new units to the system, is by changing the demand characteristics through load management and by adjusting the generation characteristics of the system. Adjusting generation characteristics, usually the unit size and the FOR, is normally a matter to be considered only when planning the installation of new units for production, and is discussed extensively in the second part of the book.

Load management can alleviate the reliability problem by changing the load shape and the load factor, usually by getting customers to shift consumption from on-peak to off-peak hours. The primary means by which load is managed include pricing policies which assign higher rate for consumption of power during peak periods, storage devices to store energy produced during off-peak hours for consumption during peak periods, and load-shedding schemes which interrupt service, in a controlled manner, during periods of capacity deficiencies. Of the three means, the latter seems to offer an off-the-shelf, cost-effective device to control load, while the others still require additional research and development before they can be widely implemented.

It should be noted that while it is assumed in the discussion that energy curtailment will be effected as soon as generation capacity falls below customers' demand, in practice it is possible for the electric power company to continue servicing all customers past this point by reducing system voltage and frequency (brownouts). While the major reason for voltage reductions is transmission inadequacy, brownouts may also be applied in cases when the reserve, on line and spinning, is insufficient to accommodate the demand for electric power. Voltage reductions generally occur in steps ranging from approximately 3% to a maximum of 8%. Reductions beyond 8% exceed the tolerance built into many appliances and other equipment and thus can cause damage. The major impact of brownouts is overheating and lower performance of motors, less light from bulbs, and similar inconveniences. The computer industry, for instance, can be seriously affected by voltage fluctuations. Consequently, such companies usually install protective devices to avoid damage and to prevent the destruction of data so as to minimize damages from brownouts. In a typical system a 5% reduction in voltage is estimated to shave some 3% off the peak, while a drop of 8% in voltage is believed to shave approximately 5% off the peak (5). Hence, there is a built-in reserve of almost 5% in most systems due to brownouts. Putting it differently, brownouts make it possible for a power company to continue service with somewhat lower quality even if generation capacity is as low as 95% of that needed for maintaining normal voltage.

If voltage or frequency reductions do not suffice to overcome capacity shortages, load-shedding devices could be effected to cut off some customers from the network, thus relieving the problem of excess demand. When load shedding is done on a controlled basis, and applied principally to loads which are insensitive to timing of supply, the LOLP experienced by the system will not be affected, as the load curtailed will not go unserved but, rather, will be switched to periods of slack demand. In principle, we can distinguish between two types of load-shedding schemes:

1. Load shedding—large customers. In cases where customers have standby facilities (large industrial customers), or are not too sensitive to the timing of their supply (large water-pumping systems), special arrangements can be made to interrupt service to those customers at specific times, perhaps by offering them a rate reduction as a benefit.
2. Load shedding—ripple or radio controlled. This relatively new technique permits widespread cutting of small customers by enabling the dispatcher to turn off many smaller loads from the dispatch center. A sensing switch is installed at certain load centers which are not sensitive to the exact timing of supply (e.g., electric water heaters). The sensor is able to identify radio command or coded pulses sent directly over the ac line. When identifying a certain code, the load is turned off; when sensing a second code, the load is turned on. When implemented on a wide basis, the method permits the shedding of a large load from the network.

If, after exhausting all the possibilities for brownouts and load shedding, capacity deficiencies still exist, there is no choice but to disconnect whole regions or groups of customers from the network, resulting in a curtailment of power on a nondiscriminatory basis. To minimize the damage to customers, a priority scheme might be utilized. Under such a scheme, customers in each region would be classified according to the extent of damage caused by power curtailment; the higher the damage, the higher the priority. Interruptions are then set according to these priorities. First, regions with the lowest priority should be disconnected. As shortages increase, regions with higher priorities will be cut off supply. In addition, regions with the same priority can be made to share power interruption through a rotating scheme. First, some regions are interrupted, while others get supply. When duration of interruption exceeds a predetermined value, the interrupted regions are reconnected to the network, while the others are disconnected. Such a scheme is somewhat analogous to the merit order loading of generators, except that here we are unloading power according to the merit order of load centers.

If these means still do not suffice to solve the capacity deficiency problem, other measures should be taken to avoid a total collapse of the system, such as automatic load shedding and disconnection of generators. All these methods, however, including the priority load-shedding scheme, result in an increase in the LOLP of the system and in poorer reliability performance.

As mentioned, the two other means of load management are pricing policies and energy-storage devices. Both are aimed at shifting load from on-peak to off-peak hours, thus affecting reliability performance through the resulting impact on demand. A combination of the three methods might ultimately be required to manage demand and control peak load growth, with great potential for improving system reliability without having to invest in extra installed capacity.

5

Operating Cost Calculation

5.1 Introduction

The prediction of the operating and the capital costs of an electric power utility is an important aspect of system planning. In capacity expansion decisions, cost data are used as a major criterion in comparing various alternative investment policies for the purpose of finding the least-cost investment which meets the demand for power with a certain level of reliability. In operation planning, cost calculations are needed for financial planning, cash-flow analysis, fuel budgeting, etc. Cost estimations are also important elements in supporting rate structure decisions.

The prediction of the capital cost of an investment plan is in itself a complicated task. It involves forecasting the capital cost of installed capacity for each type and size of generating unit in each future year. The estimation of the operating cost of an electric power system is far more complex as it depends on the loading procedure, availability of units, and the demand for electric power, which are highly variable and unpredictable, especially when the calculations extend far into the future.

Various methods have been developed for predicting the operating cost of a given utility. The simplest procedures are based on the load–duration curve (LDC) assuming a merit order loading, with the capacities of the units reduced by a few percent to account for outages and scheduled maintenance as described in chapter 3. A linear version of this estimate has been used as the cost function in various linear programming models (1). A more sophisticated approach, which accounts for forced outages and maintenance in a more reliable way, while still assuming a merit order loading, is the probabilistic simulation method (15, 19), which will be described in detail in part II. A greater degree of precision in calculating the operating costs can be achieved by using a simulation model, as the analysis can be carried out on an hourly basis and more accurate loading procedures, specifically incremental loading, can be employed. The disadvantage of simulation models, however, is the large volume of calculation involved. Also, as the analysis extends further into the future, simulation models become ineffective since it becomes almost impossible to predict hourly data that far in advance. In this chapter we show the operating cost calculation for systems without forced outages which are loaded according to incremental (marginal) cost loading. Calculations for systems with forced outages will be shown in part II.

5.2 The Incremental Loading Procedure

Assuming that units possess nondecreasing and differentiable costs functions, it can be shown (12) that the incremental loading procedure results in a minimum for the operating cost of the system. In the following the cost functions will be denoted by

$$c_i = f_i(p_{L_i}), \qquad i = 1, 2, \ldots, N, \tag{5.1}$$

where

p_{L_i} = the output (in MW) of generator i,

$f_i(p_{L_i})$ = a nondecreasing and differentiable function, expressing the cost of generating at output level p_{L_i} for one time unit,

N = the number of generators in the system.

The marginal cost function of a unit i is derived from the cost function by differentiation:

$$c_i' = f_i'(p_{L_i}), \qquad i = 1, 2, \ldots, N.$$

Let us assume for convenience that the marginal cost functions are one-to-one, so that their inverses exist and are also one-to-one. The inverse functions are denoted by

$f_i^{-1}(c_i'), \quad i = 1, 2, \ldots, N$; then

$$p_{L_i} = f_i^{-1}(c_i'), \qquad i = 1, 2, \ldots, N.$$

The corresponding cost function and marginal cost function for the whole system will be denoted by $c = f(p_L)$ and $c' = f'(p_L)$, respectively, where p_L is the total generation of the system, satisfying[1]

$$p_L = \sum_{i=1}^{N} p_{L_i}, \qquad c' = c_i', \qquad i = 1, 2, \ldots, N.$$

In the following we will ignore start-up and shutdown costs and consider fuel cost only. (Start-up and shutdown cost might be estimated

[1]It is worthwhile noting that if $c_i' = f_i'(p_{L_i})$ are viewed as supply curves, as defined in microeconomic theory, then $c' = f'(p_L)$ is the resulting aggregate supply curve and is obtained in the same manner, i.e., by adding outputs for identical marginal costs.

quite accurately from past data and can be added to the operating cost of the system as a lump sum.)

The optimal loading problem can be presented in the following manner:

$$\text{minimize} \quad \sum_{i=1}^{N} f_i(p_{\mathrm{L}_i}) \tag{5.2}$$

$$\text{subject to} \quad \sum_{i=1}^{N} p_{\mathrm{L}_i} = p_{\mathrm{L}} \qquad \text{(demand constraint)},$$

$$p_{i\,\min} \leq p_{\mathrm{L}_i} \leq p_{i\,\max} \qquad \text{(technical constraint)}.$$

Relaxing the inequality constraints for a moment, the problem can be solved by the use of the Lagrange multiplier. Differentiating the Lagrangian

$$L = \Sigma f_i(p_{\mathrm{L}_i}) - \lambda\left(\sum_{i=1}^{N} p_{\mathrm{L}_i} - p_{\mathrm{L}}\right) \tag{5.3}$$

yields the optimal strategy

$$\frac{\partial c_i}{\partial p_{\mathrm{L}_i}} = \lambda, \tag{5.4}$$

namely, the marginal operating costs of all units should be equal. The inequalities can be dealt with by assuming $c_i = 0$ for every $p_{\mathrm{L}_i} \leq p_{i\,\min}$ and $c_i = \infty$ for every $p_{\mathrm{L}_i} \geq p_{i\,\max}$.

The incremental loading procedure can be demonstrated graphically. For simplicity we assume a three-unit system with linearly increasing marginal costs c'_i, $i = 1, 2, 3$, as described in figure 5.1. The marginal cost function for the system, obtained by summing up the abscissas of the individual marginal cost functions for each marginal cost level, is also exhibited in figure 5.1. Given a demand level p_{L}, the corresponding generation levels p_{L_i}, $i = 1, 2, \ldots, N$, are determined, according to the incremental loading procedure, at the point where the marginal costs of the various generating units are equal to the marginal cost of the system. To derive these levels graphically we draw a horizontal line through $c'(p_{\mathrm{L}})$ which cuts the ordinates of the individual marginal cost curves at the points $c'_1(p_{\mathrm{L}_1}), c'_2(p_{\mathrm{L}_2}), \ldots, c'_N(p_{\mathrm{L}_N})$, respectively. The generation level of each unit p_{L_i}, $i = 1, 2, \ldots, N$, is then read directly off the abscissa. The procedure is demonstrated in figure 5.1.

When the marginal cost of the system is higher than the maximum marginal cost of a certain generator (obtained by evaluating the marginal cost function at its upper capacity), the generator is scheduled to work at its maximum output; and conversely, if the marginal cost for the system is below the minimum marginal cost of a certain generator (obtained by

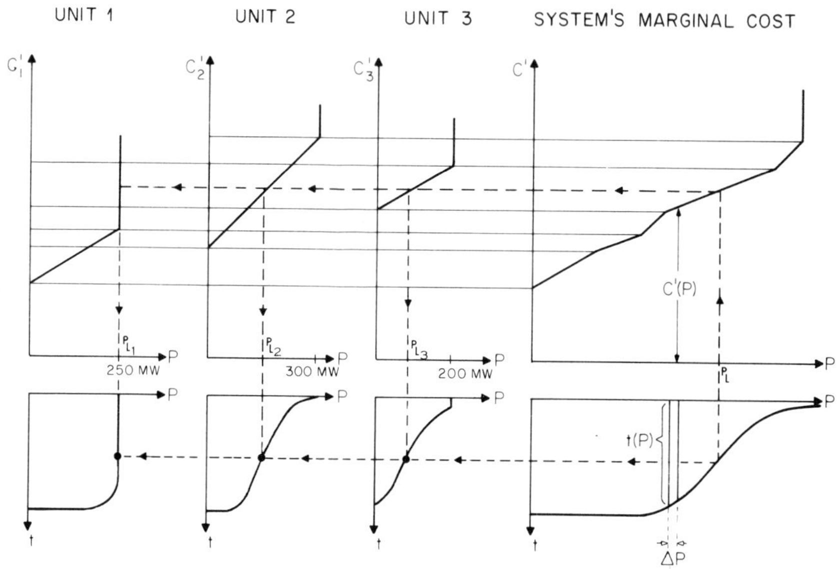

Figure 5.1 (A) Loading of three units according to incremental cost method. (B) Calculating the annual operating costs of the system. Key: $p_{i\ min} = 0$, $i = 1, 2, 3$; $p_{1\ max} = 250$, $p_{2\ max} = 300$, $p_{3\ max} = 200$.

evaluating the marginal cost function at the point of minimum production), the generator is scheduled to work at its minimum output.

5.3 Calculating the Total Operating Cost of the System

The total operating cost of the system is obtained from the system's marginal cost function and the LDC. Given a demand level greater than p_L, the incremental energy supplied to the customers at cost $c'(p_L)$ is $t(p_L)\,dp_L$, where $c'(p_L)$ is the marginal cost of the system at demand p_L. The cost to deliver this incremental energy is

$$c'(p_L)\,t(p_L)\,dp_L, \tag{5.5}$$

from which the total operating cost to the system is obtained by integration:

$$\mathrm{TC} = \int_0^{p_{max}} c'(p_L)\,t(p_L)\,dp_L, \tag{5.6}$$

where p_{max} is the peak demand for the period.

5.4 The Energy Delivered by Each Unit

Denoting the LDC for generator i by $t(p_{L_i})$, we have

$$E_i = \int_0^{p_{max}} t(p_{L_i})\, dp_{L_i}. \tag{5.7}$$

However, $t(p_{L_i})$ is not known, as it depends on the loading pattern of the units to generation, which is, in turn, a function of the demand for electric power. To find $t(p_{L_i})$ we will have to consider the contribution of generator i to total generation at all levels of the demand p_L.
Multiplying and dividing (5.7) by dp_L we have

$$E_i = \int_0^{p_{max}} t(p_{L_i}) \frac{dp_{L_i}}{dp_L}\, dp_L. \tag{5.8}$$

But at the point of equal incremental cost $t(p_{L_i}) = t(p_L)$; hence,

$$E_i = \int_0^{p_{max}} t(p_L) \frac{dp_{L_i}}{dp_L}\, dp_L. \tag{5.9}$$

To find dp_{L_i}/dp_L we make the substitution

$$\frac{dp_{L_i}}{dp_L} = \frac{dp_{L_i}/dc'}{dp_L/dc'}, \tag{5.10}$$

and since $p_{L_i} = f_i^{-1}(c')$ and $p_L = \sum_{i=1}^{N} p_{L_i}$ we obtain

$$E_i = \int_0^{p_{max}} t(p_L) \frac{(f_i^{-1}(c'))'}{\sum_{i=1}^{N} (f_i^{-1}(c'))'}\, dp_L. \tag{5.11}$$

Taking a closer look at the latter formula it can be noticed that the term in the numerator is the reciprocal of the slope of the marginal cost of generator i at a demand level p_L; the corresponding term in the denominator is the reciprocal slope of the marginal cost function for the system at demand p_L. The incremental demand of energy $t(p_L)\, dp_L$ is thus shared by the various generators in the system according to the above ratio. Then by integrating over all possible demand levels p_L, the energy delivered by each unit is obtained.

An extension for finding the LDC for each unit $t(p_{L_i})$ is immediate. Denoting the previous ratio by $m(p_L)$ we obtain from the above

$$dp_{L_i} = m(p_L)\, dp_L, \tag{5.12}$$

which upon integration yields

$$p_{L_i} = m(p_L) p_L. \tag{5.13}$$

Hence for each demand level p_L, the corresponding generation level p_{L_i} is obtained by multiplication by the ratio $m(p_L)$. Since by definition $t(p_{L_i}) = t(p_L)$, the LDC for unit i is thus obtained.

5.5 Finding Fuel Cost for Each Unit in the System

The fuel cost (or the operating cost) for each unit in the system can be found in a similar manner. Recalling that the incremental energy cost for unit i is given by

$$dc_i^* = c_i'(p_{L_i})\, t(p_{L_i})\, dp_{L_i}, \tag{5.14}$$

it follows that

$$c_i^* = \int_0^{p_{max}} c_i'(p_{L_i})\, t(p_{L_i})\, dp_{L_i}. \tag{5.15}$$

Substituting $t(p_{L_i}) = t(p_L)$, and since by the incremental loading procedure $c_i'(p_{L_i}) = c'(p_L)$, we have

$$c_i^* = \int_0^{p_{max}} c'(p_L)\, t(p_L)\, dp_{L_i}. \tag{5.16}$$

Multiplying and dividing by dp_L we have, as before,

$$c_i^* = \int_0^{p_{max}} c'(p_L)\, t(p_L)\, m(p_L)\, dp_L. \tag{5.17}$$

The incremental cost $c'(p_L)\, t(p_L)\, dp_L$ is again shared by the various units according to the reciprocal ratio of the slope of their marginal cost value at point p_L to the slope of the marginal cost function of the system at that point. Note that the energy delivered by each unit in the system can be obtained as a special case of cost calculations by substituting $c'(p_L) = 1$ for all levels of p_L in equation (5.17).

Given the operating cost for unit i, and the unit cost of fuel, the amount of fuel consumed by each unit in the system can be determined.

6

The Use of Quantitative Methods in Short- and Long-Term Planning of Installed Electric Energy Generating Capacities

6.1 Introduction

Electric power plays an important role in the economic development and well being of nations. The consumption of electric energy generally increases, with an annual growth rate varying roughly in the range of 3 to 20%, depending mainly on the economic growth and industrialization rate of the particular nation. Power generating systems are highly capital intensive, this characteristic becoming more dominant as a result of increase in units' size, new technologies, and automation. Thus, investments involved in developing and expanding such systems to meet rising demands amount to many billions of dollars, even for small nations.

The vast investments needed, which present much more of a problem to developing nations, call for efforts to use existing quantitative tools, and forge new ones when necessary, as effective aids in determining optimal, or at least near-optimal, plans for development and expansion of power generating systems.

The optimal mix problem was one of the first problems considered in the analysis of power systems (4, 9, 33). Given an array of plants which differ from one another in the initial cost, expected useful life, and running cost per kilowatt-hour, an optimal mix which will minimize the total cost of the system for a given load–duration curve is sought. Because of the complexity of the capacity expansion problem, the optimal mix problem was solved initially only for stationary systems. Subsequently, mathematical programming was employed for solving the problem for dynamic systems.

The general, vaguely defined problem of devising an optimal development plan for the total system is very intricate, involving many factors—in particular, political and social factors—not amenable to quantification. In view of the impracticality of attempting to tackle the general problem with quantitative tools, two major simplifying assumptions are made here, leading to a problem twice removed from the general one, but still quite realistic in nature.

First we assume that future demand is given and must be satisfied, to a certain well-defined degree, with minimum cost. This assumption means that demand is an exogenous force not affected by interaction with availability of supply. Accordingly, forecasts of future demands are derived on the basis of past trends, population and economic growth projections, and the like (using various forecasting methods). Under this assumption the planner's major decision variables are

1. the types of generating facilities to be installed (e.g., nuclear, conventional oil-fired, gas turbines, etc.);
2. the size and introduction time of each unit;
3. the geographic location of each unit; and
4. the transmission network development.

A second simplifying step is to disregard decision variables (3) and (4), which is, in a way, equivalent to assuming that all generating units are located at one place and all demand is concentrated at one point. This simplification is made since this book deals only with generating systems, and dealing with variables (3) and (4) would involve a lengthy discussion.

The problem is now restricted to determining the optimal types, sizes, and future introduction times of power generating units, so as to satisfy a projected future demand to a certain defined degree (expressed in terms of expected number of days with shedded load, loss-of-load probability, or a similar measure).

Viewing the problem with the planner's practical objectives in mind, one is led to make a distinction between short-term and long-term planning. Long-term planning must provide orders of magnitude and a general framework, leaving enough freedom of action for adapting to possible changes of circumstances and, at the same time, providing sufficient information for making the necessary preparations of a rather general character for meeting demands in the distant future. Short-range planning, on the other hand, is much more of a commitment and must be detailed and specific. A period of about 10 years seems appropriate for short-range planning. Long-range planning is constrained by the decreasing reliability of forecasted information and 30 years seems a reasonable limit in this respect.

It is worthwhile pointing out that in most cases the planner is forced to bring into consideration existing commitments at the starting point.

The question of what tools we should use to tackle the short-range and the long-range planning problems is answered by the nature of these plans. As mentioned, short-range planning requires accuracy and detailing, necessitating consideration of stochastic elements (forced outages), weekly, or even daily, changes in load–duration curves, detailed maintenance schedules, and, above all, consideration of each unit individually. The tool most appropriate for treating a problem of this type is digital simulation. In contrast, the requirements of long-range planning allow aggregations of units, use of yearly load–duration curves, and use of averages for representing forced outages and maintenance schedules. Hence the problem can be posed as a deterministic one and tackled by an appropriate mathematical program. The size of the problem (constraints and variables), however, dictates the use of a linear program.

Two very common models are presented here for the sake of demonstration, a simulation model for the short range and a linear program for the long range. There are numerous possible models for both the short and the long run. These two were selected due to their popularity.

It is important to point out that the simulation program does not yield a particular optimal solution; it merely evaluates any plan that is fed into the program. The choice among the alternatives is made outside of the simulation, and the program supplies all the data regarding costs and operation of a given alternative.

6.2 The Traditional Optimal Mix Problem

Assume that (a) the capital (investment) cost per unit of installed capacity is constant for any given type of unit regardless of its size (i.e., no economies of scale), (b) the running costs per unit of energy output (kW h) are fixed for all output levels, (c) the plant capacity is measured by its mean availability, and (d) the capital and operating costs for the various types of generating units are inversely related. Under these conditions, the optimal mix is found by using an analysis of the break-even point of the total leveled annual capital charge and the running costs while comparing each two competing units. First the optimal number of running hours is determined for each given type of unit; then the installed capacities are determined. The optimal running duration of each type is found as follows:

Let

p_j = the capacity of type $j, j = 1, \ldots, J$, ordered according to their merit order,

$K_j = \sum_{i=1}^{j} p_i$,

F_j = annual fixed costs of type j (per kW),

V_j = variable costs of type j (per kW h).

Then total costs of the system will be given by

$$\mathrm{TC} = \sum_{j=1}^{J} F_j(K_j - K_{j-1}) + \sum_{j=1}^{J} \left(V_j \int_{K_{j-1}}^{K_j} T(x)\,dx \right). \tag{6.1}$$

Its minimum is obtained by solving

$$\frac{\partial \mathrm{TC}}{\partial K_j} = F_j - F_{j+1} + V_j T(K_j^*) - V_{j+1} T(K_j^*) = 0. \tag{6.2}$$

We get the optimal operating duration for type j as

$$T(K_j^*) = \frac{F_j - F_{j+1}}{V_{j+1} - V_j}. \tag{6.3}$$

Knowing the optimal duration, we can obtain the optimal installed capacity directly from the LDC (figure 6.1). The system's LDC is used to determine the optimal capacity of each plant. We note that practically, since capacity has to be added in discrete increments, the system might have at any time more or less of a particular type of capacity than an optimal system would indicate. (It simply never makes sense to add just 1 kW.) However, by assuming away economies of scale, one can infer that the excess capacity can be rented or sold in the short term, so that the net system, adding purchases and subtracting sales, may be close to optimal, at least under the stationarity assumptions stated above.

The primary drawbacks of this optimal mix approach stem from the underlying assumptions, which usually do not hold in reality. For example, due to economies of scale, both the capital cost per kilowatt and the running cost per kilowatt-hour vary for any given type of plant. Being a model for finding the optimal mix at a given time, no account is taken of the interaction between the decision variables over time. Also, the procedure for approximating forced outages and maintenance requirements using the plant's average availability is indeed very coarse.

The deterministic methods based on the LDC, which ignores the forced outages rate, fail to describe accurately the system performance at peak hours as they never expect that peaking units, which are more expensive in operation, will ever be used. This is because any system, if it is to be at all reliable, has an excess installed capacity over demand; thus, by assumption, peaking units will never be scheduled for operation as there will always be a cheaper unit around to run. This drawback of the deterministic methods stems from the fact that they usually do not account for forced outages and maintenance, which have a substantial impact on the performance of a power system, especially during peak hours. An attempt to correct this fault by reducing the capacity of the generating units by a few percent in order to account for the fact that power plants are not available 100% of their time is equivalent to assuming an even distribution of maintenance and forced outages over the period, an assumption that usually does not hold in reality. Phillips et al. (25) have suggested an extension to the optimal mix approach to account for the interaction of the operating cost effects between different years by setting up the problem as a series of related 1-year problems, but they run into serious computational difficulties. No satisfactory mathematical programming procedure is

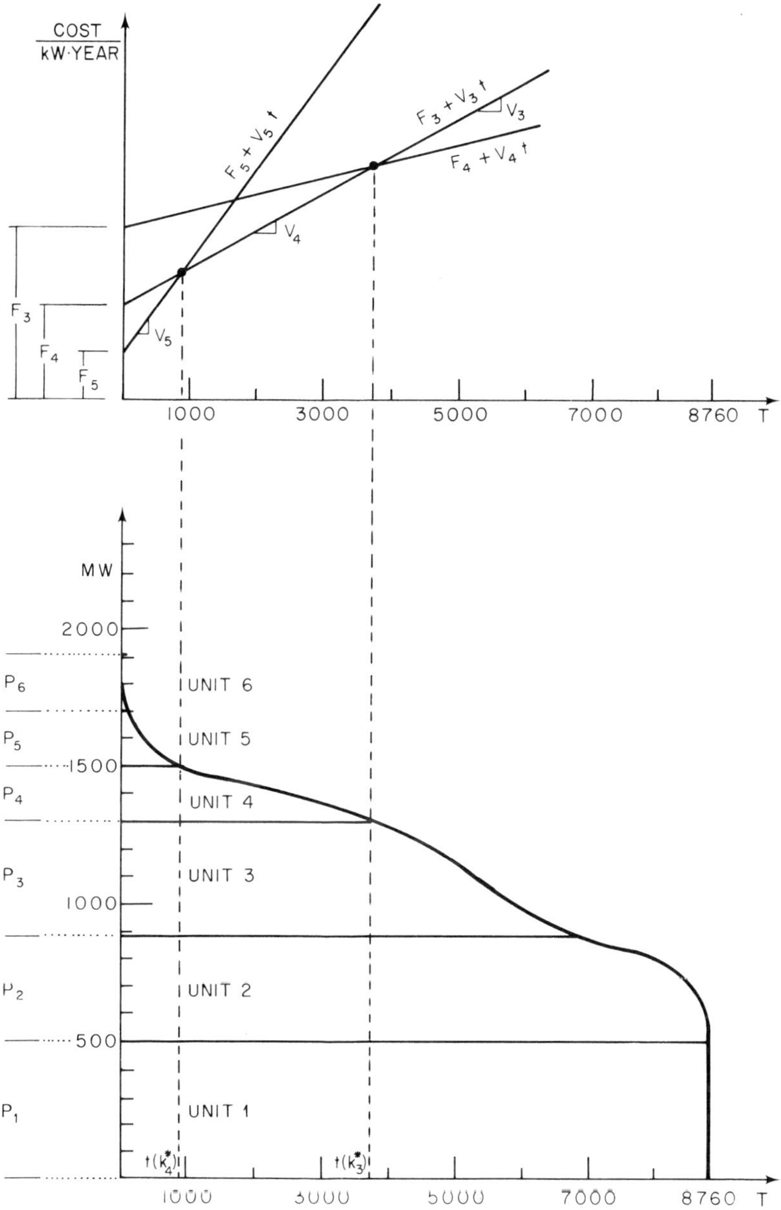

Figure 6.1 Determining generating mix by break-even approach. Example shows six types of units. Calculations for types 3, 4, and 5 are shown in detail. p_j is the optimal installed capacity of type j, $j = 1, 2, \ldots, 6$.

available yet for accounting for forced outages and maintenance requirements in a more accurate manner than the mean availability approach.

Consequently, the stationary model has very little value for planning purposes, and can serve at best to yield a first approximation to the optimal mix of a system at a particular point in time. For small systems that experience only a small increase in demand for power this approximation may give fairly good results.

While the assumptions underlying the optimal mix problems are indeed very restrictive for base and intermediate units, they may be more realistic when applied to peaking units. The fixed and variable costs of peaking facilities do not vary substantially with the size of the unit, and their unconditional outage probability (not to be mixed with FOR as explained before), considering that peaking units work only a small fraction of the time during a year, approaches zero. Given the mix of base-load and intermediate-load units, one can use this model to find the corresponding optimal mix between various peaking facilities with a reasonable degree of accuracy.

6.3 The Linear Programming Model

The linear programming (LP) model described is based on D. Anderson's (1) model, which was slightly modified to better cope with peaking units in light of their high energy costs. The load–duration curve used by this model is the equivalent LDC to be discussed in detail in part II. In order to understand this chapter it is enough to realize that the curve here better reflects unit loading, especially peaking units. The regular LDC can be used and the interested reader is referred to the original formulation by Anderson.

For deriving the LP model we first represent the projected annual equivalent LDCs by approximate block diagrams. This is illustrated in figure 6.2. The blocks are numbered from left to right using the index $n = 1, 2, \ldots, N$. The width of the nth block is denoted by θ_n and is the same for all the years of the long-range plan, $t = 1, 2, \ldots, T$. The height of the nth block is different for each year and is denoted by Q_{nt}. The approximate area under the equivalent LDC of year t is given by $\sum_n \theta_n Q_{nt}$.

In the following we shall use the index $j = 1, 2, \ldots, J$ to denote units of type j, and the index v to denote the vintage (year of introduction) of the units. Our decision variables are of two types.

1. The total capacity, in megawatts of type j units introduced in the year v. This variable is denoted by x_{jv}.
2. The utilization of type j and vintage v capacity, during year t ($t \geq v$), in block n of the combined LDC. This variable is denoted by u_{jvtn} and may

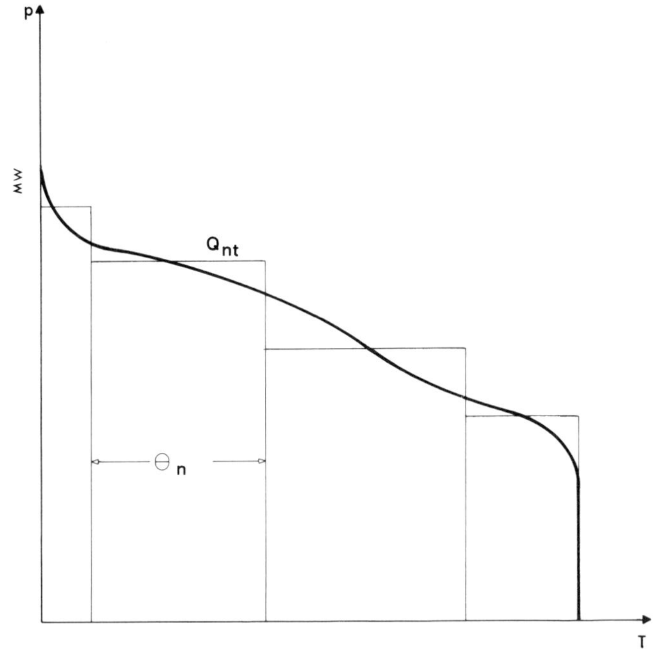

Figure 6.2 Approximating the equivalent LDC by a block diagram.

be better understood by the observation that the contribution of x_{jv} to the area of block n of year t is equal to $\theta_n u_{jvtn}$, $t \geq v$.

Since utilization cannot exceed capacity, we have

$$u_{jvtn} \leq x_{jv}, \qquad \begin{aligned} j &= 1, 2, \ldots, J, \\ n &= 1, 2, \ldots, N, \\ v &= v_0, v_0 + 1, \ldots, T, \\ t &= v, v + 1, \ldots, T. \end{aligned} \tag{6.4}$$

The assumption that all demand must be met at all times[1] is expressed by

$$\sum_j \sum_{v \leq t} u_{jvtn} \geq Q_{tn}, \qquad \begin{aligned} t &= 1, 2, \ldots, T, \\ n &= 1, 2, \ldots, N. \end{aligned} \tag{6.5}$$

[1]Observe that it is possible to increase or decrease the amount of unsatisfied demand by changing the value of Q_{t1}. Another method for introducing some level of shortage will be discussed later.

Our objective is to satisfy constraints (6.4) and (6.5) at minimum cost. Thus, the function to be minimized is

$$w = \sum_{j} \sum_{v} c_{jv} x_{jv} + \sum_{j} \sum_{v} \sum_{t} \sum_{n} f_{jvt} \theta_n u_{jvtn} a_{jvt}, \tag{6.6}$$

where

1. c_{jv} is the present value of the fixed cost per megawatt of type j capacity introduced in the year v.
2. f_{jvt} is the present value of the utilization cost of generating a megawatt-hour during year t by type j units of vintage v.
3. a_{jvt} is a correcting factor representing the estimated proportion of real power generation. The area under an equivalent LDC is comprised of energy demand (real demand) and forced outages and maintenance demand. The factor a_{jvt} can be viewed as the proportion of time during which the units produce power. A trial-and-error may be needed to adjust the assumed values of these factors with model results. This adjustment may be eliminated when using the Anderson model. It will, however, lead to less accurate results, expecially in peaking units.

The majority of the constraints of the LP model given by relations (6.4), (6.5), and (6.6) are the capacity constraints described by relation (6.4). For units whose capacities are known (e.g., the units that are already in existance) the x's are constants, and the capacity constraints become simply upper bounds on the corresponding u variables, so they do not affect the number of rows of the simplex tableau. Where the x's are variables, it is possible to decrease significantly the number of capacity constraints by noting that equivalent LDCs are monotonically decreasing and as a result the order $u_{jvt1} \geq u_{jvt2} \geq \cdots \geq u_{jvtN}$ must hold in an optimal solution. Defining new values

$$\begin{aligned} z_{jvtn} &= u_{jvtn} - u_{jvt(n+1)}, \qquad n = 1, 2, \ldots, N-1, \\ z_{jvtN} &= u_{jvtN}, \qquad u_{jvtN+1} = 0, \end{aligned} \tag{6.7}$$

it is now possible to replace constraints (6.4) by relations (6.7) and the following constraints:

$$\sum_{n=1}^{N} z_{jvtn} = u_{jvt1} \leq x_{jv}, \qquad \begin{aligned} j &= 1, 2, \ldots, J, \\ v &= v_0, v_0 + 1, \ldots, T, \\ t &= v, v + 1, \ldots, T. \end{aligned} \tag{6.8}$$

This transformation increases the number of nonzero coefficients, but it also decreases the number of rows. It is best to use this z formulation for those plants whose x's are unknown, but to keep the u formulation with bounded u variables for those x's that are fixed.

At first glance it seems that the cost function given by equation (6.6) is not realistic since it assumes that the fixed cost per megawatt c_{jv} and the utilization cost per megawatt-hour f_{jvt} are independent of the number of units introduced at year v and of their sizes. It can be shown that this is true for gas turbine units only, and this is due to the assumption that all such units are of the same standard size. Still, equation (6.6) is quite a good approximation. When the sizes of the new units to be introduced during the long range vary between 600 and 1200 MW, a linear cost can be assumed. The problem of the utilization cost f_{jvt} is solved by iterations. When starting, sizes of newly introduced units during various periods of the long range should be assumed, using the results as the new assumptions for the next run until convergence is achieved.

The end of the planning period can have a significant effect on the later years of the plan. If a simple cutoff is made, the model's solution is incorrect. The solution would state that in the later years one should build plants with low-investment cost, even if utilization costs of production are very high. For example, the model would require many gas turbines. It would not be reasonable to build nuclear plants for only a few years of usage.

There are several ways of avoiding this problem. First, the model can be run for a number of years after the time period of interest and the results for the last years can be ignored. Along with this, the last, or two last, time periods can be made of longer duration. Alternatively, the costs of the u variables can be increased for the last time period, to give that period greater weight. This is equivalent to running an additional number of years with constant demand.

The planning model presented is also intended to admit the possibility of not satisfying all of the demand all of the time, but to permit some level of shortages. For this purpose, we include one or more fictitious units with $c_{jv} = 0$ and with high f_{jvt} value. This f_{jvt} value represents the penalty to the economy for not supplying a unit of required energy. For any estimate of this penalty, the program can be used to determine the optimal solution and the associated level of shortages.

The LP model can be run for a large number of alternative conditions with the purpose of investigating the influences of possible changes in key factors (such as projected demands, fuel costs, investments, etc.) on the long-range plan. The output obtained from the different runs is rather extensive.

6.4 The Mixed Integer Programming Model

Another fault of the linear model is that the numerical values obtained as solutions for x_{jv} have to be rounded off to the nearest standard unit sizes.

An almost ideal representation of the long-range planning problem could be achieved by using mixed integer programming. With this method, in addition to the decision variable x_{jv}, which represents the total amount of capacity, we can define integer variables I_{jvk} representing the number of units of type j, vintage v, and size k, which are to be introduced. To insure the correct value of x_{jv} we would require that

$$x_{jv} = \sum_{k} k\, I_{jvk} \tag{6.9}$$

for all suitable j and v. The cost coefficient of I_{jvk} would be the unit's total cost.

Since the capability of mixed integer programming codes is limited, it is necessary to limit the number of I variables. They should be chosen to represent the most pertinent decision variables, e.g., for nuclear plants, or for the years for which decisions must be made immediately.

6.5 The Simulation Model

The simulation model simulates the operation of the system on a weekly basis. It accepts as input a description of the current system and of the investment alternatives. Also, it reads the necessary data that enable it to develop a load–duration curve for each week of the planning period. Then, for each week, after scheduling maintenance and randomly generating breakdowns according to stipulated distributions, it schedules the supplying of the demand by the operating plants in sequence of their loading procedure. The loading procedure can be either merit order or incremental cost. For the sake of simplification the model presented here will use merit order loading, though the incremental loading procedure will yield more accurate results. The program accumulates statistics regarding costs and other relevant factors. The required program input includes

1. a description of the relevant characteristics of each unit that exists at the beginning of the planning, and of each new unit that is scheduled to be introduced during the planning period (the relevant characteristics are size, efficiency merit order, fixed costs, fuel consumption, availability date, forced outages probability, and maintenance requirements);
2. a projected weekly demand (kW h) for the duration of the planning period;
3. descriptions of several typical weekly LDC shapes with an indication of which shape is appropriate for each week of the year.

After accepting this information the program builds the expected LDCs for all the weeks of the planning period and sequentially schedules the weekly supplying of power. For each week, only those units that are operative are used, taking into consideration the maintenance schedule and the possibility of forced outages. The available units are assigned according to their merit order, namely, the most efficient unit is assigned first, etc. The output includes weekly, yearly, and cumulative costs, energy production of all units, load (plant) factors, distribution of shortages, and some other relevant data.

Since the simulation is performed for a relatively short period (about 10 to 15 years) a problem exists with regard to the proper evaluation of the residual values of the system at the completion of the period.

Apart from using the simulation model for evaluation of alternative short-range plans, it is used as an aid in projecting the equivalent load–duration curves used later in the LP model. This curve differs from the regular curve in that it shows not only the customer demand but statistically adds to it the capacity which is under forced outages and the capacity under maintenance, thus yielding a much more realistic picture of units loading. A detailed description of the simulation model is given in the appendix.

6.6 Discussion—The Use of Planning Models

The short-range simulation model and the long-range LP model are of complementary nature and should be used accordingly. One major problem of the simulation model is the difficulty of evaluating the residual value at the end of the planning period. This problem does not exist when the two models are used in a complementary fashion, since the short-range plan is actually the initial condition for the long-range one.

The methods and programs are general in the sense that they can easily be adapted as useful tools in the planning of various electric power generating systems.

The projected annual energy demands (areas under the annual load–duration curves) and peak demands are used in building complete forecasted load–duration curves (LDCs) which serve as input for the long-range planning LP. For the short-range planning simulation model one has to build weekly projected LDCs. The task of building these projections involves laborious data collection and analyses, and takes a good deal of time and effort. The description of the methodology used for developing the needed projections, which no doubt may be of interest to some readers, exceeds the limited scope of this book.

In view of the complexity and magnitude of the generation planning problem it is doubtful that without certain simplifications any decision model, however elaborate, can yield an optimal solution to the capacity expansion problem. The best that can be hoped for is a near-optimal solution. Even when a system attains an economic equilibrium, changing conditions, such as availability of various types of fuel, fluctuations in fuel

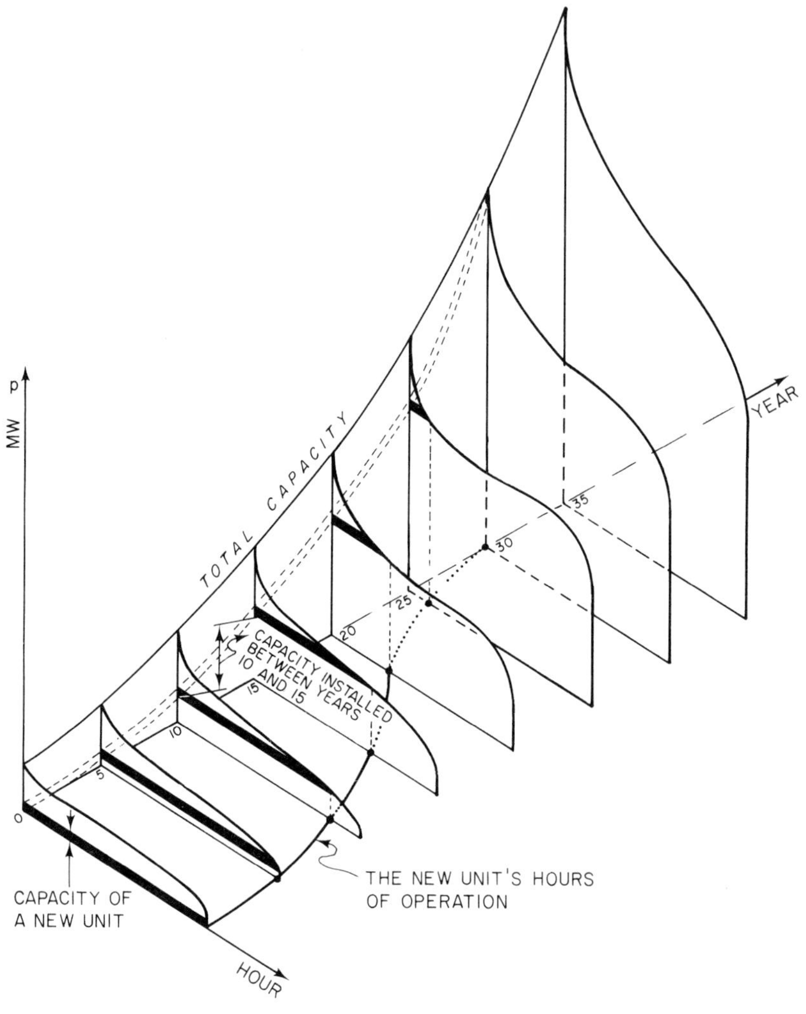

Figure 6.3 Generating unit move from base-load to peak-load service during its lifetime.

and construction prices, and new environmental regulations, tend to shift the system away from optimality. Fortunately, because of the dynamic characteristics of the system, and the need for continually expanding the system to meet the expected increase in demand, the expansion program of a power system can be self-corrected. For example, in an electric system base units are moving upward as time passes because of the introduction of more efficient units (figure 6.3). Therefore, if, for instance, in a given year a system contains more base units than is economically justified, short-term expansion plans can favor investments in peaking units, thus shifting the system toward optimality; and, on the contrary, if a system has too many peaking units, future expansion plans can emphasize investments in base and intermediate units, thus correcting the system. Hence, if adjusted year-by-year, it is possible to drive a system toward optimality, even when the current system is not optimal.

7

Rate Structure Analysis—Allocation of Marginal Costs under Supply Uncertainties

7.1 Introduction to the Marginal Cost Approach to Rate Making

The economic prescription for attaining an optimal allocation of resources is the equation of price and marginal cost. *Marginal cost* is the cost of producing one more unit, or, alternatively, the cost that can be saved by producing one less unit, of an item. In an economy with a limited amount of resources, any increase in consumption for a given commodity will obviously result in a decrease of production of all other commodities. The cost to society of producing an additional unit of a given item is, therefore, equal to the opportunity cost of producing less of the other commodities. If buyers are charged less than the marginal cost of producing the item, they will tend to expand their consumption; society will have to sacrifice more of other goods and services in order to meet the increased demand for the product. Conversely, if prices are set above the incremental cost, consumers will buy less of the commodity than is economically justified, resulting in unnecessary scarcity. The economic equilibrium is achieved, therefore, when prices are equal to the marginal cost (20).

As long as our goal is economic efficiency, the notion that prices should be equal to marginal cost is a general economic principle, applying also to electricity. The situation in the electric power industry, however, is more complicated. Because the demand for power is periodic, varying by time of day and season of the year, and its supply basically nonstorable, we find that the marginal cost of supplying additional consumption also varies. Consumption of power at peak hours requires that we run high energy-cost peaking units more intensively, while off-peak consumption can be met by operating cheaper-energy plants at a much lower cost. Consequently, the marginal cost of meeting the demand during peak hours is considerably higher than the marginal cost for providing electricity during off-peak hours. If this variation in marginal cost is to be reflected in the rate structure, prices of electricity should be set at different levels for different hours and seasons. Various schemes have been devised for pricing electricity, including flat rates, straight-line meter rates, declining block rates, rates with demand components, incremental cost rates, inverted rate structure, and peak-load pricing. Rate structure analysis is heavily dependent upon intensive cost studies (both capital and operating costs) and careful estimation of the short- and long-run marginal cost of generating electricity. This is by no means an easy task, because of the changing pattern of demand, the variability in the units available for production at any point in time, fluctuations of construction and fuel costs, the variety of generating units in a modern utility, etc. Methods for calculating operating costs of power systems have been described in the literature. Various techniques for estimating capital and marginal costs are available (24).

Pricing systems based on marginal cost which vary by the time of day and season of the year have generally been referred to as *time-of-the-day–seasonal pricing*, or *peak-load pricing*.

Peak- (variable) load pricing has been gaining a good deal of popularity in recent years as an approach to marginal cost allocation. In principle, peak-load pricing is the application of the principles of marginal cost pricing to a situation in which the marginal costs exhibit a variable pattern. Some of the basic principles of peak- load pricing have been described in a collection of articles by French economists edited by Nelson in 1964 (24). Kahn in 1970 (20) further elaborated on peak-load pricing aspects and peak-responsibility principles. Recent testimony by National Economic Research Associates (NERA) (16, 22, 31) summarizes some practical aspects of peak-load pricing and suggests methods for assigning marginal costs to various customers. The history and the current activity of peak-load pricing is also contained in a recent paper by Joskow (17). While the notion itself is not new, it has drawn much attention recently as a possible means of managing demand and controlling peak-load growth, primarily because of the energy crisis and the associated increase in fuel prices and construction costs of power plants.

While theoretically well defined, the task of translating the economic principle into actual pricing schemes is not that simple. Beyond difficulties in measuring marginal and opportunity costs, there are other qualifications, discussed widely in the economic literature (20), that should also be accounted for by the price scheme. Among these qualifications are income distributions, equity considerations, second-best factors, financial aspects, and other social goals. Additional constraints are imposed by the need for simple tariffs, risk, ignorance about consumer reactions, and technical difficulties relating to metering and billing. Yet, whatever modifications are necessary to account for these additional concerns, the starting point of any price scheme must still be the marginal-cost structure, in order to ensure that any deviation of prices from the incremental cost will have the least damage on the allocation of resources.

7.2 Allocation of Marginal Costs under Supply Uncertainties

A further complication, beyond the time-dependent variability of marginal costs, is introduced into marginal pricing of electricity because of the random fluctuations of the installed capacity due to forced outages, maintenance requirements, and the random deviations of customers' demand, rendering the marginal costs random variables themselves.

Taking uncertainty of supply and demand into account, Balasko in 1974 (2) verified the well-known principle that an optimal pricing rule should equate the price of electricity to the expected energy (operating) cost plus the expected marginal curtailment cost. Crew and Kleindorfer (10) have established the same conclusion by formulating the problem of determining optimal pricing and capacity policies as a welfare maximization problem while accounting for demand uncertainties as well as the diversity of modern power generation means.

The two major subjects of rate structure analysis according to the marginal cost approach are: how to allocate the marginal capacity cost and how to allocate the marginal energy cost among the different customers. The effects of the stochastic nature of the electric generation system must be taken into account in calculating and allocating the marginal costs.

7.3 Marginal Capacity Cost under Supply Uncertainties

An issue of particular controversy in peak-load pricing has been the problem of charging customers with capacity costs. It is usually conceded that a sustained increase in off-peak consumption might be met in the long run by changing the mix of plants that we have on line, to take advantage of the generation economies associated with higher load factors, without having actually to increase the overall capacity of the system. And conversely, a sustained increase in on-peak consumption calls for adding extra capacity to the system, or otherwise the incremental on-peak demand will not be met. Consequently, according to these traditional methods, on-peak customers are the only customers who cause the expansion of the system and should therefore assume full responsibility for incurring capacity costs. As Kahn (20) puts it, "Every purchase at that (consumption of power at peak) time makes its proportionate contribution in the long run to the incurrence of these capacity costs, and should therefore have that responsibility reflected in its price. No part of these costs as such should be levied on off-peak users" (p. 89).

This traditional conclusion has recently been challenged by a few writers. Wenders, in a paper published in 1976 (38), shows that, contrary to the above result, capacity costs will also appear in the prices of "intermediate" and even off-peak periods. The testimony by NERA advocates the use of the loss-of-load probability (LOLP) in assigning marginal capacity costs to various periods. Streiter (31) has even provided a simplified example, with made-up curtailment probabilities, to demonstrate the calculation procedure.

In this book the theoretical and empirical approach for charging off-peak customers with capacity costs is further developed. It will be shown that the problem of who should bear the responsibility for capacity costs actually boils down to the design target of the system, because it is the attainment of this design criterion that eventually calls, in the long run, for adding extra capacity to the system (36). The problem of apportioning the marginal costs among the various customers, therefore, derives from this observation. Thus, in those power utilities in which the installed capacity is set equal to the peak demand, or to the peak demand plus a certain percentage of it, on-peak customers are obviously the only contributors to the need to add capacity to the system, and should therefore assume full responsibility for these capacity costs. However, modern power utilities are not designed just to meet the peak demand for power, but, rather, to deliver power at a certain level of reliability. Then, since the reliability performance of a power system is affected by all types of customers, it follows that a sustained increase in demand at any hour, including off-peak hours, calls for adding extra capacity to the system, or otherwise the reliability design target will not be met. Consequently, each hour contributes its own share to the need to incur capacity cost, and should therefore have that responsibility reflected in its price.

In apportioning marginal capacity costs, it is therefore necessary to study first the impact of each hour on the attainment of the system's design target, and then to allocate the capacity cost appropriately. Taking the LOLP design target, the contribution of each hour to the LOLP is established by studying the impact of changing the load at any hour on the system's reliability performance. Then an analytical algorithm to determine accurately these contributions is devised, and a procedure to allocate the marginal capacity cost to various customers based on these contributions is suggested. The rigor of the analysis is extended by suggesting provisions for incorporating maintenance requirements and random deviations of customers' demand in the calculation process.

It is very important to note the relation between capacity costs and the design target of the system. While the current literature is confined to the LOLP design target only, the algorithm for apportioning capacity costs will be extended to power systems designed to meet loss-of-energy probability (LOEP) reliability design targets.

7.4 Marginal Energy Cost under Supply Uncertainties

Marginal energy costs over time constitute a major component of schemes of peak load pricing. In power systems without forced outages, the marginal energy cost associated with each hour of consumption is merely the energy cost of the last unit loaded. However, when a system is exposed to forced outages, various units might assume the role of the last unit at any given hour, rendering the marginal energy costs random variables.

The computation of the expected marginal cost is complicated by the stochastic nature of the availability of plants, the loading pattern of units to production, the maintenance schedule, which affects the units that might be available for production at any point in time, and the randomness of the demand for power. Consequently, there is no efficient direct procedure as yet for ascertaining the expected marginal energy costs for each hour of consumption, even for the simplified case of merit order loading.

By definition, the marginal energy cost for each hour is the cost of providing an additional unit of output, i.e., the cost of an additional kilowatt-hour. In a merit order loading, an additional increment in demand at any hour will be picked up by the last unit in the merit order which is scheduled to operate at that hour, or, if the last unit is loaded to full capacity, by the next available unit in the merit order. The marginal energy cost associated with the hour is therefore equal to the average production expenses of this last unit. For example, in figure 3.5A we describe the merit order loading for a power system, with unit 1 denoting the most efficient unit, unit 2 the second most efficient unit, and so on. Assuming all units are in working condition, an increment of demand at, say, hour t will be picked up by unit 3, at a cost of C_3 ¢/kW h, so that the marginal cost of hour t for this particular case is C_3—the average production cost of unit 3. From here on we will refer to the last unit called upon for generation at any given hour as the *last unit* or the *last supplying unit* for the hour. (This should not be confused with the last, or the most recent, unit installed in the system.) Obviously, each unit can assume the role of the last unit, depending upon the availability of units and the demand for power.

In power systems with no forced outages, it is quite easy to calculate the marginal energy cost associated with each hour, since it is equal to the average production cost of the corresponding last unit. For example, if figure 3.5A describes such a system, the marginal energy cost of hour t is constant and equal to C_3 ¢/kWh. However, in power systems in which the generating units are subject to forced outages, there is a certain probability that each of the different units will be assigned to generation as the last unit for any given hour, depending on different combinations of units in and out, each resulting in a different marginal energy cost. Going back to figure 3.5, suppose that unit 1 is down because of a forced outage. Then all the other units will occupy a lower position in the merit order, as shown in figure 3.5B, with unit 5 now serving as the last supplying unit for hour t. Hence the marginal energy cost for hour t, for this realization, will be C_5 ¢/kWh, equal to the average production cost of unit 5. Assuming the standard two-state model for each generating unit due to forced outages (operative and thus capable of full power delivery, or nonoperative and incapable of producing any electricity), as many as 2^N realizations are possible, each with its own probability and each resulting in a different time-dependent distribution of marginal energy costs. The expected value of the cost of providing an additional kilowatt-hour at any given hour is therefore obtained by multiplying the various marginal costs by their corresponding occurrence probabilities and summing up the weighted costs over all possible realizations. Using the enumeration approach, the calculation process consists of the following steps:

1. Generate the capacity table of all combinations (events or realizations) of up and down units along with their occurrence probabilities (see table 3.1).
2. Identify, for each combination, the range of hours served by each working unit.
3. For each combination, calculate the weighted marginal energy cost of each hour by multiplying the average fuel cost of the unit which provides the electricity at that hour by the occurrence probability of the event.
4. Sum up the weighted costs, for each hour, over all possible realizations, to yield the expected marginal energy cost for the hour.

The main drawback of this procedure is its tendency to result in an excessive amount of calculations, which increase exponentially with the number of units in the system.

A way to approximate power supply uncertainty in calculating the cost of meeting an increment of energy is to use the effective capacity approach in conjunction with merit order loading. We recall that according to this approach, forced outages and maintenance requirements are taken into account by reducing the rated capacity of each unit by a certain factor, reflecting the average fraction of time the generating unit is not available for production. The marginal energy cost associated with a given hour is then equal to the average production cost of the unit which provides the energy at that hour, with the units loaded to generation, in the merit order, at their effective rather than their rated capacity. While the approach is indeed very simple, it does not account for maintenance and forced outages requirements in a realistic manner since it assumes an even distribution of outages over the year. As already mentioned, it has been verified that the effective capacity approach tends to underestimate the production expenses of a power utility (28, 29). Prices based on marginal energy costs calculated by this approach are likely to be underestimated.

Simulation methods provide alternative means for accurately accounting for supply and demand uncertainties in calculating expected marginal energy cost for each hour of consumption. However, simulation models are very expensive to run and are not flexible enough for performing various needed analyses as part of the calculations.

Consequently, a more efficient, preferably analytical, procedure for calculating data on expected marginal energy costs is required to support decisions regarding peak-load pricing. Such a procedure, which is presented in the second part of the book, results in substantial savings in the amount of calculations when compared with simulation, let alone direct enumeration, methods. The idea will be to derive demand-related expected marginal energy costs first, and then to attribute each cost to the corresponding hour of the year by using the load–duration curve of the system.

II

Considering the Random Nature of Electric Energy Generating Systems

The first part of this book was aimed at acquainting the reader with the central issues of planning electric power generating systems and analyzing their short- and long-run performance. This was done through an extensive overview of the prevalent basic concepts, such as system reliability, reserve, and costs, and the prevailing traditional methods of decision support analyses.

This part of the book is concerned with more recently developed methods and techniques for tackling the problems and issues considered in the first seven chapters, and uses a unified approach based on the equivalent load–duration curve (ELDC). We will start with the method for deriving the ELDC and proceed to develop its uses in computing reliability measures, operating costs, and capital costs, and in relating these quantities to each other in a formal and precise manner. The stochastic nature of the supply of electric power, brought about by forced outages, plays a dominant role in the development of these methods and techniques.

The heart of the approach which will be described and implemented is the equivalent load–duration curve. In order to formulate the curve, the demand for power is extended to include four components: the deterministic component of customers' demand; the random component of customers' demand; "demand" for forced outages; and "demand" for maintenance. Interpreting the latter two "demands" as the self-demand of the electric company for installed capacity, the sum of these four components, referred to as the *combined*, or the *equivalent*, *demand*, dictates the total requirements for generating capacity needed for satisfying customers' demand and maintenance and forced outages requirements. Under certain assumptions, the complementary cumulative density function of the combined demand is obtained by means of a convolution process. This function, which is the ELDC, relates the combined demand for power to the expected proportion of time in a given period, usually a year, that each load level will be equaled or exceeded. Given an explicit form of the LDC, a computational procedure for calculating the ELDC is derived directly from the convolution formula, resulting in substantial savings in the amount of computations. The significance and uses of the

ELDC are then described. Unlike the regular LDC, the ELDC is a function of both demand and generation characteristics of the system. Thus, any change in the parameters of the system is reflected in the ELDC.

While the existing methods for supporting planning decisions in electric power systems are drawn from various disciplines, including mathematical programming, simulation, economic analysis, marginal analysis, and probability theory, the ELDC provides a unified approach to supporting the whole range of decision problems related to planning in the electric power field (42). The application of the ELDC to reliability analysis, sensitivity and trade-off analysis, reserve planning, operating cost calculations, rate structure decisions, and capacity expansion programs is the essence of this book.

The wide spectrum of applications provided by the ELDC renders it a most powerful and versatile tool for the analysis of power systems. The benefits of having a unified multipurpose tool to support planning decisions, rather than many single-purpose methods, are (a) it reduces the efforts expended in the modeling process, (b) it yields consistent results, (c) it provides more insight into the system by consistently identifying the complex interactions inherent in system operation and expansion, and (d) it consolidates the decision-making process.

8

The Equivalent Load–Duration Curve (ELDC)

8.1 Introduction

The need for better analytical tools that will account in a more accurate manner for the random nature of electric systems due to forced outages has resulted in the development of a modified version of the LDC, known as the *equivalent load–duration curve* (ELDC). The equivalent load curve was first initiated by Baleriaux et al. (3), and was later refined by Booth (9), Joy and Jenkins (19), and others. In its most general form, the ELDC contains four elements of demand (9): the deterministic component of customers' load; the random component of customers' load; requirements for forced outages; and requirements for maintenance. Figure 8.1 shows the actual load distributions of a specific utility including maintenance and forced outages. Joy and Jenkins (19) have demonstrated how the outage probabilities are convolved with the load density function to yield an equivalent load curve of customers' demand and forced outages. By using a transformation formula from probability theory, they were able to derive the equivalent load curve in a recursive manner. The resulting curve has then been applied to calculating the expected operating cost of a power system, the amount of energy delivered by each unit, the LOLP, and the loss-of-energy probability.

An alternative computational procedure for computing the ELDC, derived by describing it as a convolution, is presented in this chapter.

8.2 Definition and Interpretation of the Equivalent Load–Duration Curve

In the LDC the demand for power is identified with customers' load only. An alternative approach is to look on the demand for power in terms of all the factors which "compete" for the generation capacity of the system. These are the customers' demand as well as requirements for outages and maintenance.

Figure 8.2 presents two ways to look at the system. According to the traditional one presented in figure 8.2A, shortages, marked by darker

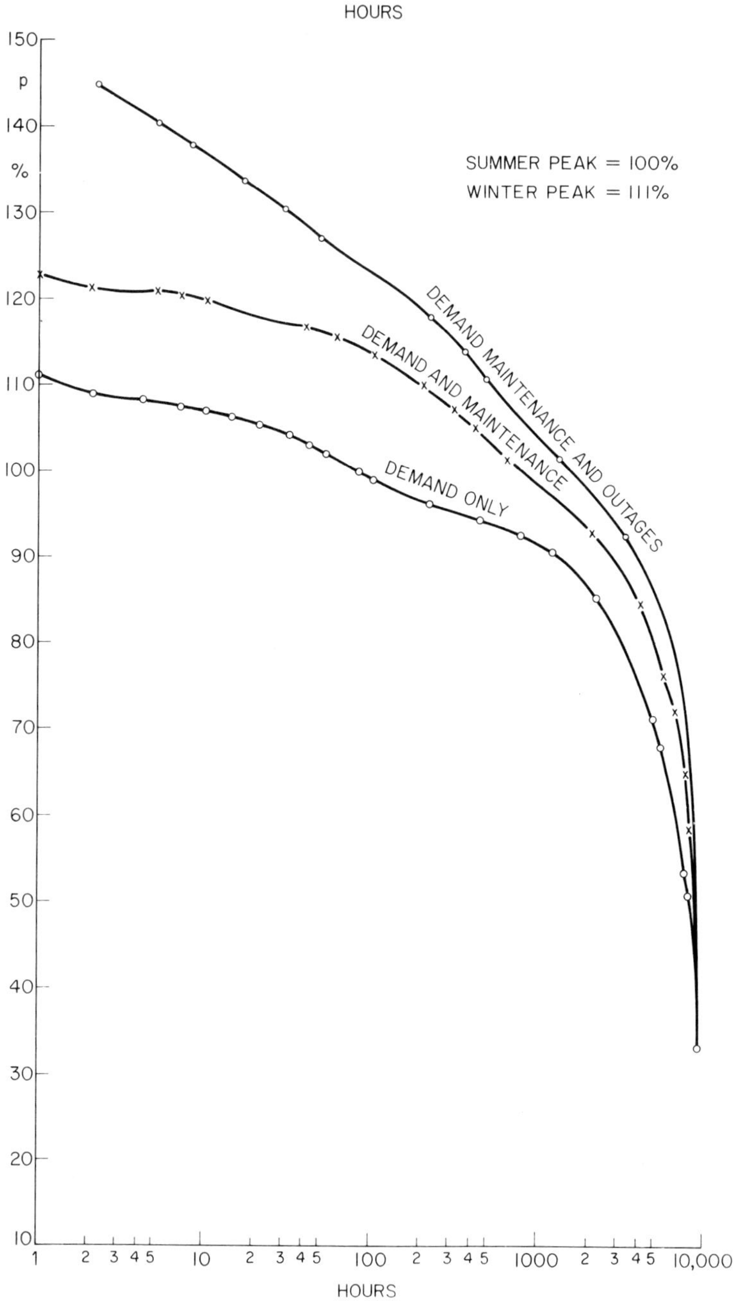

Figure 8.1 Actual distributions of system load plus capacity under maintenance and forced outages in a given year for a specific utility.

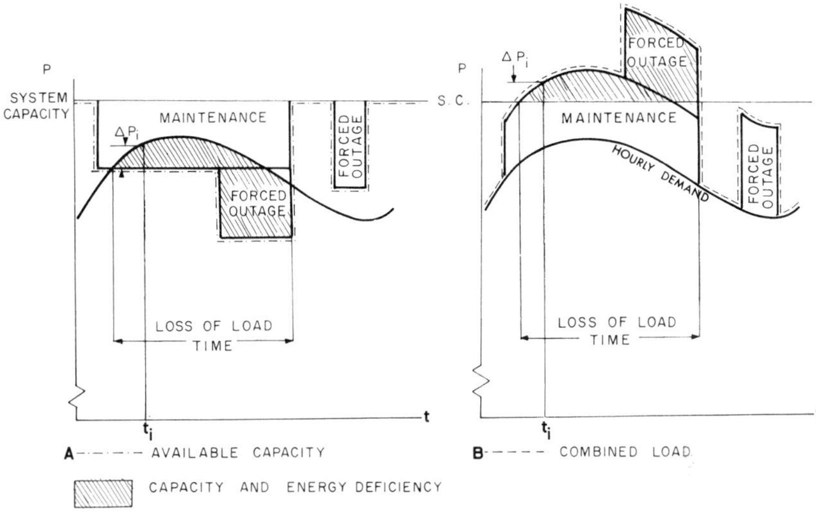

Figure 8.2 Two ways of representing the margin between the demand and the available capacity.

areas, result from a discrepency between the demand and available capacity, where the available capacity is the installed capacity less the units under maintenance and forced outages. Looking at figure 8.2B will reveal, however, that the very same result is achieved by assuming constant system capacity and treating the capacity under maintenance and forced outages as an extra load. The resulting shortages are identical.

The equivalent (or combined) demand is defined as

$$p_C = p_D + p_R + p_M + p_F, \tag{8.1}$$

where

p_C = the combined demand,

p_D = the deterministic component of load,

p_R = the random component of load,

p_M = "demand" for maintenance,

p_F = "demand" for forced outages.

For better comprehension, the "demand" for forced outages and the "demand" for maintenance can be conceived of as the self-demand of the electric system with zero consumption of energy. The deterministic component of load can be conceived of as the expected customers' load, with the random component p_R distributed about it, usually in a normal manner, as described in figure 3.3. The combined load p_C specifies what portion of the installed capacity is occupied in satisfying customers' demand or engaged in maintenance or forced outages. Noting the correspondence with the LDC, the equivalent load–duration curve ELDC is described by $\bar{T}(p_C)$, which is the number of hours during which the combined demand is equal to, or in excess of, p_C. The normalized ELDC is described by $\bar{t}(p_C)$, which is the proportion of time during which p_C is equaled or exceeded by the combined demand. When the ELDC is given by $\bar{t}(p_C)$, a multiplication by the number of hours in the period (8760 for a year) will yield $\bar{T}(p_C)$, expressing the ELDC in terms of hours rather than proportions of time. $\bar{t}(p_C)$ can be viewed as the complementary cumulative density function of the equivalent (combined) demand:

$$\bar{t}(p_C) = 1 - F(p_C), \tag{8.2}$$

where $F(p_C)$ is the cumulative density function CDN of the combined demand, from which the combined load density function $f(p_C)$ is obtained by differentiation:

$$f(p_C) = -\bar{t}'(p_C). \tag{8.3}$$

Ignoring the maintenance for the time being, this density function can be obtained by convolving the load density function (defined in equation (3.2)) with the forced outages probabilities. A typical shape of the resulting ELDC is shown in figure 8.3.

8.3 The Probability Distributions of the Demand Components

Before deriving the convolution formula, it is necessary to define the probability distributions of its components. The load density function has already been discussed in chapter 3. The density function of the random component of customers' load is assumed to be known or else can be derived from past data. Maintenance requirements are scheduled events and are thus known in advance; the probability distribution of the forced outages can be calculated as was shown in chapters 3 and 4.

In deriving the ELDC, and for that matter in the analysis to follow, we clearly distinguish between the forced outage rate of a unit, denoted by q, and its unconditional outage probability, to be denoted by q^*. While the FOR is actually the probability that the unit will be forced out of service while at work, q^* is the probability that the unit will be forced out of service at any random point in time. By definition, the unconditional outage probability is obtained by multiplying the FOR by the fraction of time the unit is either down because of a malfunction or in service, relative to the total duration of time involved (usually a year) excluding maintenance time. Hence, while the FOR of a peaking unit is considered to be

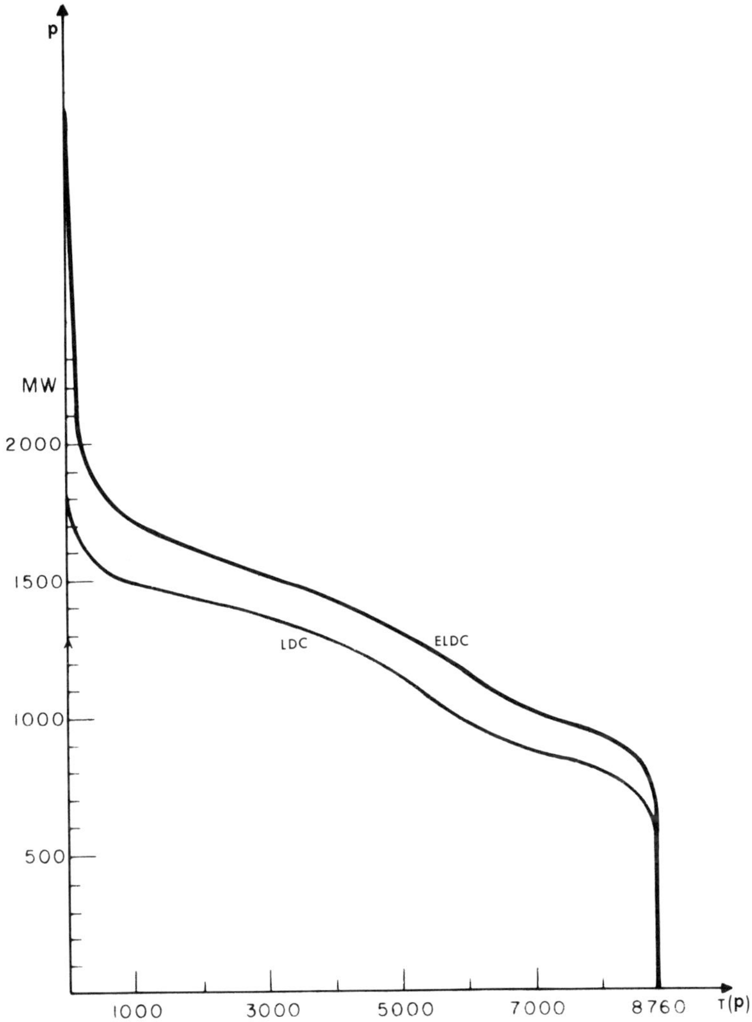

Figure 8.3 A typical equivalent duration curve for demand and outages.

quite high, its unconditional outage probability is very low, and for those peaking units which work only a very small fraction of the time during the year, it in fact approaches zero. The larger the fraction of time a unit is scheduled for operation, the closer is its FOR value to the value of q^*. As a result, the value of q^*, unlike the FOR, also reflects the location of the unit in the system's merit order, as this position determines the fraction of time the unit is scheduled for operation in a given period. A change in the unit's merit order, due to the introduction of new units, or retirement of old ones, will therefore cause change in the q^* value of the unit, while it will not necessarily affect its FOR. All formulas developed in the following are based on the FOR, unless otherwise specified. The terms FOR and outage probability will be used interchangeably, having the same meaning, and denoted by q.

8.4 Understanding the Convolution Process

The convolution process can be more easily understood by superimposing the various demand components one on top of the other while accounting for the occurrence frequencies of the events. The procedure can be described graphically for a twofold convolution only. In figure 8.4 the random component of demand is superimposed on the LDC, which is interpreted as representing the deterministic component of customers' load. The graph on the $p_C p_C$-plane defines all the combinations leading to a p_C level of combined demand, along with their probability densities. Integration of all these densities will yield the density for p_C. In figure 8.5 forced outages are superimposed on the LDC in a hypothetical system containing three units with capacities 400 MW, 300 MW, 250 MW, and outage probabilities 0.06, 0.04, and 0.02, respectively (see example, table 3.1). Eight different combinations of up and down units are possible in this case, each of which is shown in the graph along with its probability of occurrence. The lowest curve in the graph is the LDC, which corresponds to the case for which all units are operating. As can be seen in the graph, a given level p_C is produced by various combinations of forced outages and customer load. Figure 8.6 shows the weekly peaks of a system's load when maintenance is added to them, and figure 8.7 shows the resulting combined load–duration curve. The order in which the variables are convolved to yield the ELDC is immaterial; any order will lead to the same result.

8.5 Mathematical Formulation

Mathematically, the convolution formula for each period in which the same units are under maintenance during the whole period is derived as follows.

For a given deterministic demand p_D whose conplementary CDN is $t(p_D)$, the conditional density that the combined demand is at a level p_C is

$$\sum_{S} P(p_F/M)\,P(p_R = p_C - p_D - p_M - p_F), \tag{8.4}$$

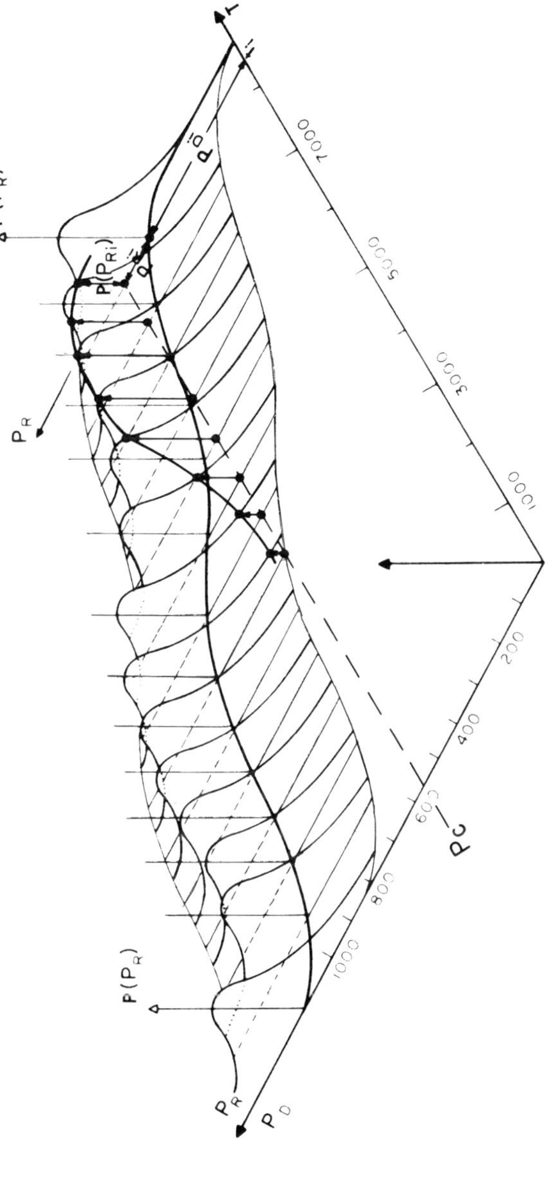

Figure 8.4 Finding the duration curve for combined deterministic plus random customers' demand.

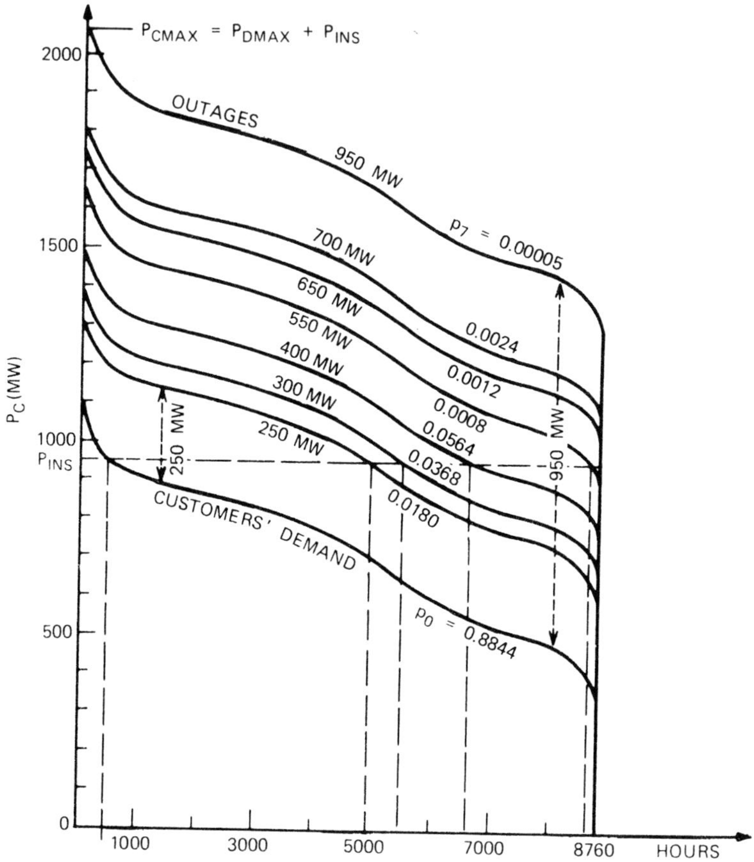

Figure 8.5 Customers' demand plus different levels of forced-outages "demand" with their probabilities.

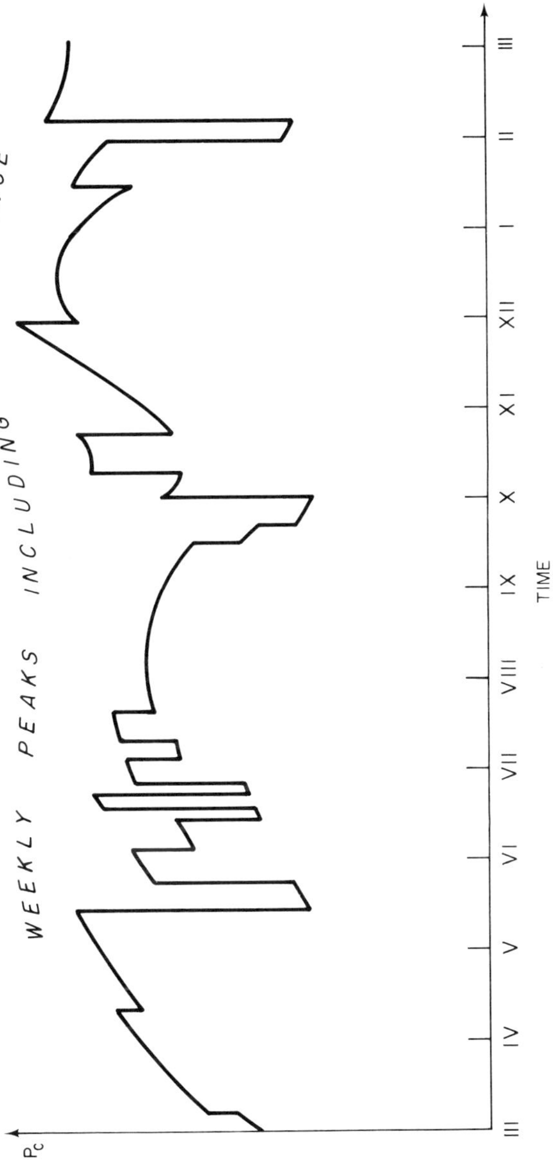

Figure 8.6 System's chronologically combined peaks including maintenance "demand."

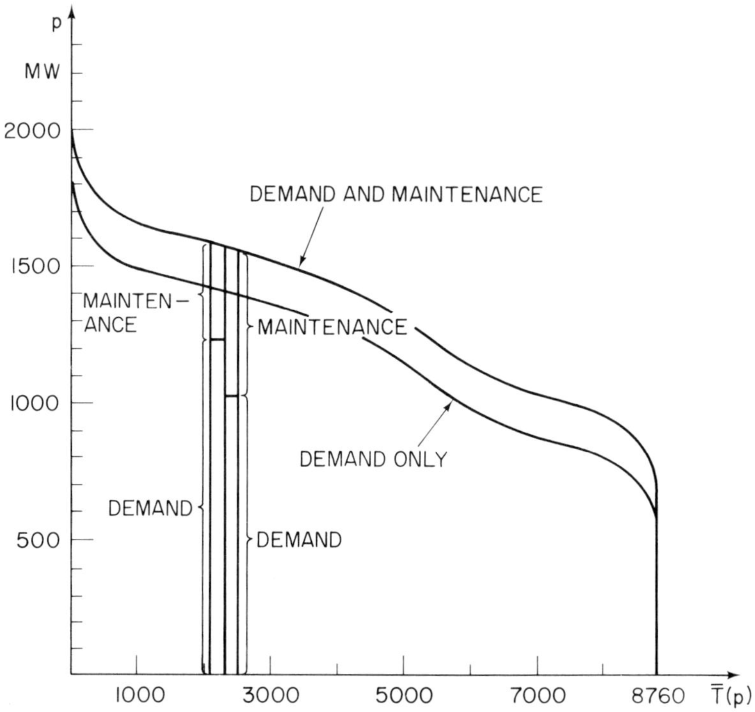

Figure 8.7 Duration curve for combined demand and maintenance.

where

$P(p_F/M)$ = the probability of outages of size p_F for the given maintenance schedule M,

p_M = total unchanged capacity under maintenance during the period,

S = the set of all combinations of up and down (due to forced outages) units for the given maintenance schedule,

$P(p_R)$ = the density of the random component of demand.

Since the density function of p_D is, as previously shown, equal to $-t'(p_D)$, the unconditional density of the combined demand obtained is

$$-\bar{t}'(p_C) = \int_{p_D=0}^{p_{D\max}} -t'(p_D) \sum_{S} P(p_F/M)\, P(p_R = p_C - p_D - p_M)\, dp_D, \qquad (8.5)$$

where $p_{\mathrm{D\,max}}$ denotes the peak level of the deterministic demand during the period.

Substituting

$$p_{\mathrm{D}} = p_{\mathrm{C}} - p_{\mathrm{M}} - p_{\mathrm{F}} - p_{\mathrm{R}},$$

and recalling that by definition the ELDC is actually the complementary cumulative density function of the combined demand, we have

$$\bar{t}(p_{\mathrm{C}}) = \int_{x=p_{\mathrm{C}}}^{\infty} \int_{p_{\mathrm{D}}=0}^{p_{\mathrm{D\,max}}} -t'(p_n) \sum_{\mathrm{S}} P(p_{\mathrm{F}}/M)\, P(p_{\mathrm{R}} = x - p_{\mathrm{M}} - p_{\mathrm{F}} - p_{\mathrm{D}})\, dp_{\mathrm{D}}\, dx. \tag{8.6}$$

Since by definition

$$t(p_{\mathrm{C}}) = \int_{x=p_{\mathrm{C}}}^{\infty} -t'(x)\, dx,$$

by interchanging the order of integration we obtain

$$\bar{t}(p_{\mathrm{C}}) = \int_{p_{\mathrm{D}}=0}^{p_{\mathrm{D\,max}}} t(p_{\mathrm{D}}) \sum_{\mathrm{S}} P(p_{\mathrm{F}}/M)\, P(p_{\mathrm{R}} = p_{\mathrm{C}} - p_{\mathrm{M}} - p_{\mathrm{F}} - p_{\mathrm{D}})\, dp_{\mathrm{D}}, \tag{8.7}$$

which is the convolution formula for the period.

The annual ELDC can now be obtained by summing over all periods of the year during which maintenance is constant, as follows:

$$\bar{t}(p_{\mathrm{C}}) = \sum_{i} \bar{t}_i(p_{\mathrm{C}}), \tag{8.8}$$

where $\bar{t}_i(p_{\mathrm{C}})$ is the ELDC for period i of constant maintenance. A block diagram of the process is exhibited in figure 8.8.

A typical ELDC and its relation to the LDC is exhibited in figure 8.3. It is important to note that while the ELDC is more or less parallel to the LDC at off-peak hours, it exhibits a shape of a "needle" during peak hours. This is because high levels of the combined demand, representing situations of high demand combined with many outages, are associated with very small and rapidly decreasing occurrence probabilities. This characteristic of the ELDC shape will prove to be of importance for various decision problems.

It is important to point out that the preceding mathematical treatment regards the hourly demands as though their random components are statistically independent. This assumption may not be realistic. Nevertheless, it may have its advantages over the approach for which the random component is ignored. An alternative approach is to assume that the LDC already accounts for both p_{D} and p_{R}. This is a special case of equation (8.7).

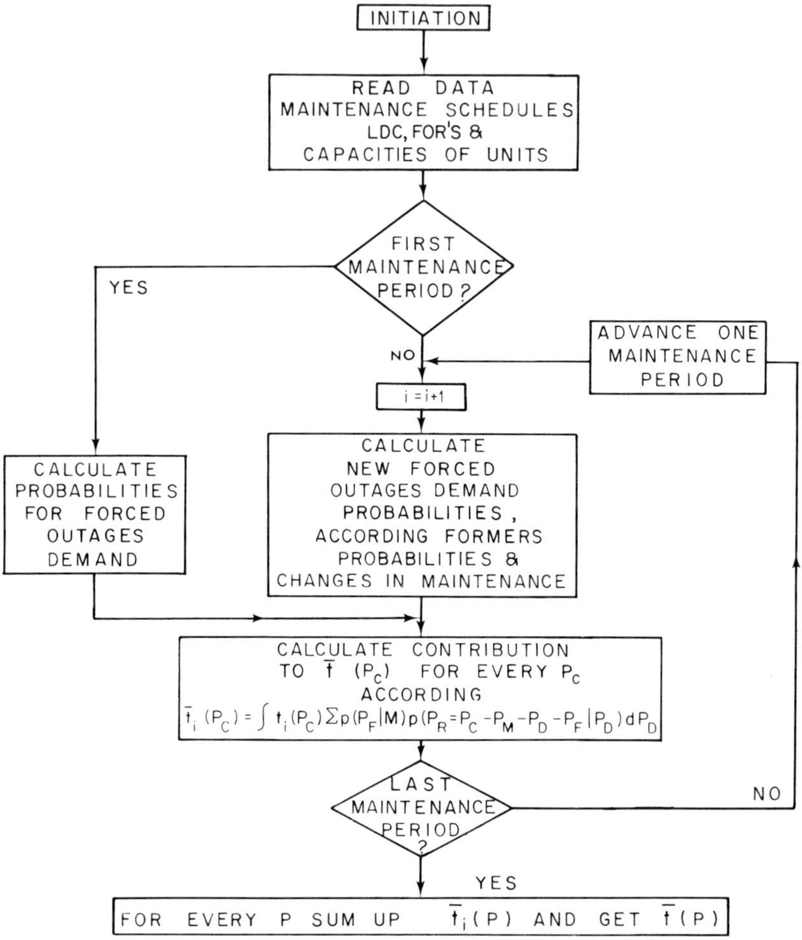

Figure 8.8 Flow chart for finding the equivalent LDC.

8.6 Incorporating Maintenance Requirements

While independent of customers' load, the forced outages probabilities are strongly dependent on the maintenance schedule. Each maintenance schedule leads to different combinations of up and down units, and hence to different failure probabilities. In order to account for this dependence the maintenance requirements were incorporated into the calculations by partitioning the period involved into subperiods of constant maintenance, and carrying out the calculations for each subperiod separately. The results of the various subperiods were then combined to yield the corresponding results for the whole period. The same procedure applies to the addition of new units to the system. Whenever a new unit is added a new subperiod must start.

8.7 Computational Aspects of the ELDC and the Recursive Equation

In order to derive the ELDC using the convolution formula there is a need to carry out the calculations over all elements in the set S, i.e., all combinations of operative and nonoperative (due to forced outages) units. The set S contains as many as 2^N combinations, with N being the number of units in the system which are not scheduled for maintenance in the subperiod. This number increases exponentially with the number of units available for production, resulting in formidable amounts of calculations. Fortunately, the task of calculating the ELDC can be greatly facilitated by using a recursive approach which was first proposed by Joy and Jenkins to define the ELDC (19). We will derive the recursive relation directly from the convolution equation.

The key to the recursive procedure is a transformation formula which enables us to calculate the ELDC of a system augmented by one unit in terms of the previous ELDC and the characteristics of the new unit, i.e.,

$$\hat{t}(p_C) = (1 - q)\bar{t}(p_C) + q\bar{t}(p_C - p_p), \tag{8.9}$$

where

$\bar{t}(p_C)$ = the ELDC prior to adding the new unit to the system,

$\hat{t}(p_C)$ = the new ELDC after augmenting the system with a new unit, of capacity p_p and outage probability q.

Proof An alternative way of expressing the ELDC is by interpreting it as a convolution of the probabilities of demand for forced outages and a partial ELDC containing the other three types of demand, i.e., the deterministic component of demand, the random component of demand, and maintenance requirements. Denoting this partial ELDC by $\bar{t}_0\,(p_C)$, we have

$$\bar{t}(p_C) = \sum_{S} P(p_F)\bar{t}_0(p_C - p_F), \tag{8.10}$$

where S is the set of all combinations of up and down units satisfying $p_F \leq p_C$.

Augmenting the system with a new unit with capacity p_p and outage probability q, the new ELDC can be obtained:

$$\hat{t}(p_C) = \hat{P} \sum_{S^*} \hat{P}(p_F) < \bar{t}_0(p_C - p_F), \tag{8.11}$$

where S^* is the new set of combinations satisfying $p_F \leq p_C$ and $\hat{P}(p_F)$ the new outage probabilities.

Under the assumption that the generating unit failures are independent of one another, the forced outages probabilities of the augmented system can be calculated using the following relationship (9):

$$\hat{P}(p_F) = (1-q)P(p_F) + qP(p_F - p_p), \tag{8.12}$$

where $\hat{P}(p_F)$, $P(p_F)$, q, and p_p are as defined above.

Substituting the latter relationship into equation (8.11) we obtain

$$\hat{t}(p_C) = (1-q)\sum_{S} P(p_F)\bar{t}_0(p_C - p_F) + q\sum_{S'} P(p_F - p_p)\bar{t}_0(p_C - p_F). \tag{8.13}$$

Note that in the first term on the right the summation is carried out over the set S, as the new unit is assumed to be operative. In the second term on the right-hand side the summation is carried out over the set S' containing all up and down combinations of old units satisfying $p_F \leq p_C - p_D$. This second term represents the case where the new unit is assumed to be down.

Substituting the expression for $\bar{t}(p_C)$ from equation (8.10), we have

$$\hat{t}(p_C) = (1-q)\bar{t}(p_C) + q\sum_{S'} P(p_F - p_p)\bar{t}_0(p_C - p_p - (p_F - p_p)). \tag{8.14}$$

By definition, the second term on the right-hand side of this equation is equal to $\bar{t}(p_C - p_p)$.

Thus

$$\hat{t}(p_C) = (1-q)\bar{t}(p_C) + q\bar{t}(p_C - p_p), \tag{8.15}$$

as required.

If in the period discussed the new unit is scheduled for maintenance, then

$$\hat{t}(p_C) = \bar{t}(p_C - p_p). \tag{8.16}$$

Figure 8.9 demonstrates the above procedure for the example system of three units when a new 25-MW unit is added to the system. Figure 8.10 exhibits the formula graphically.

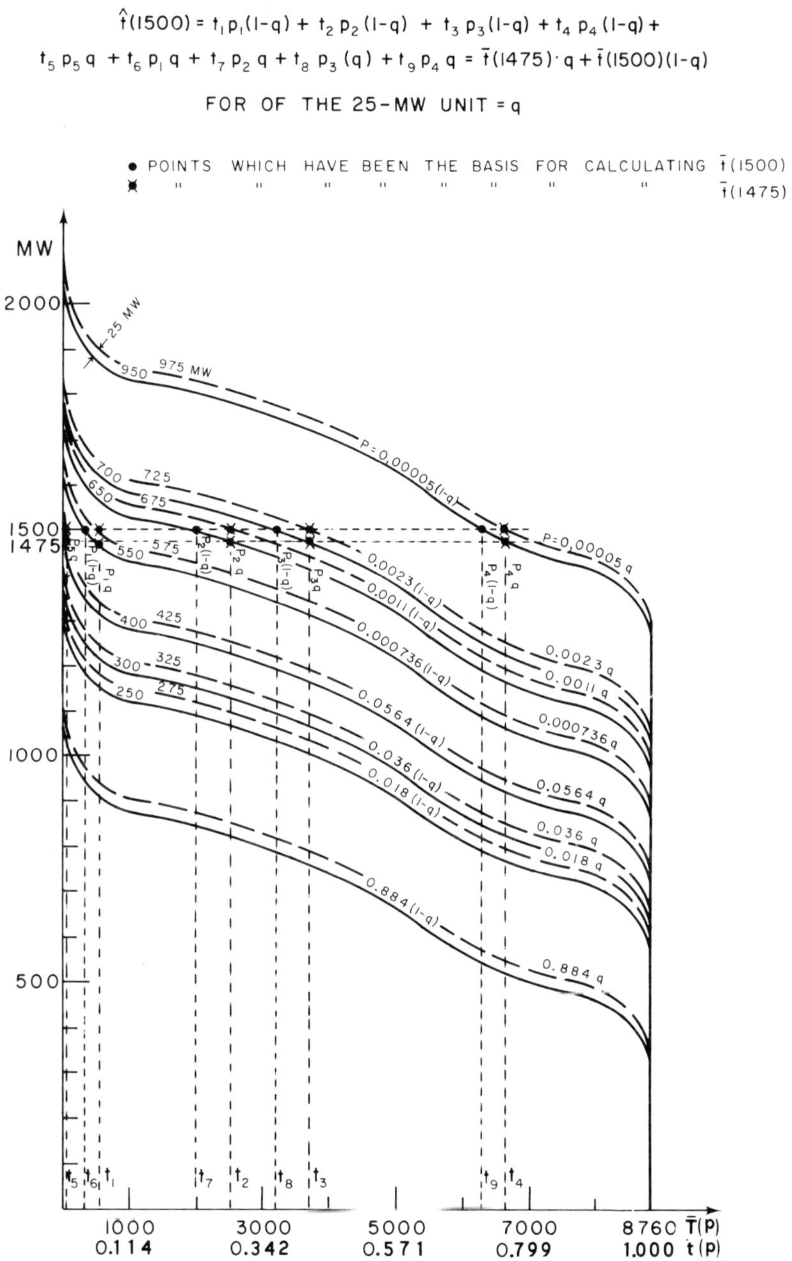

Figure 8.9 The effect of adding 25-MW unit on the equivalent LDC.

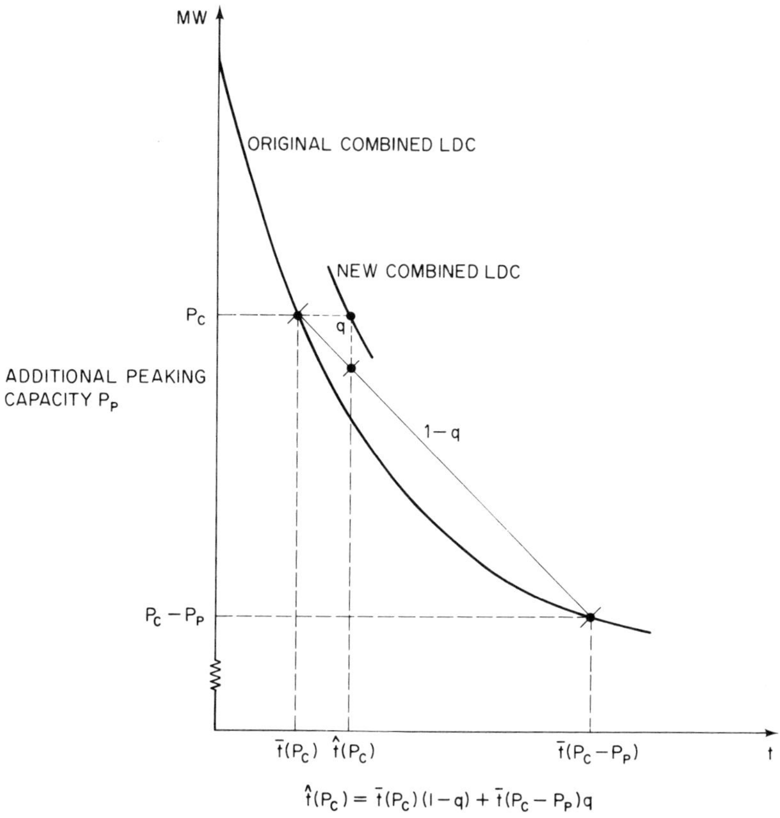

Figure 8.10 Finding the new equivalent LDC when adding a new unit with FOR $= q$.

8.8 Computing the ELDC by the Recursive Procedure

Using the transformation formula, the ELDC can be derived by adding one unit at a time in a recursive manner, until all the units in the system are incorporated into the curve.

Let

$\bar{t}_0(p_C)$ = the partial ELDC containing the customers' demand for power (both deterministic and random) and the demand for maintenance (which is fixed for each subperiod of constant maintenance),

$\bar{t}_1(p_C)$ = the partial ELDC including $\bar{t}_0(p_C)$ and the "demand" for outages of the first unit,

$\bar{t}_2(p_C)$ = the partial ELDC including $\bar{t}_1(p_C)$ and the "demand" for outages of the second unit.

In general,

$\bar{t}_n(p_C)$ = the partial ELDC including $\bar{t}_{n-1}(p_C)$ and the "demand" for outages of the nth unit.

By applying the transformation formula, the recursive relation is easily obtained:

$$\bar{t}_n(p_C) = (1 - q_n)\,\bar{t}_{n-1}(p_C) + q_n \bar{t}_{n-1}(p_C - p_n), \tag{8.17}$$

where p_n is the generating capacity (in MW) of the nth unit, q_n is its outage probability, and $\bar{t}_0(p_C)$ is given by

$$\bar{t}_0(p_C) = \int_{p_D=0}^{p_{D\max}} t(p_D)\,P(p_R = p_C - p_D - p_M)\,dp_D. \tag{8.18}$$

To obtain the ELDC for each subperiod, the process is initiated by calculating $\bar{t}_0(p_C)$, which is easily derived since it involves an evaluation of only one integral for each level of p_C, and then applying the recursive relation N times. Compared with the 2^N terms involved when applying the convolution formula, the above procedure results in a substantial reduction in the amount of calculations.

Besides its central role as a way to derive information resulting from changes in the system's generation characteristics, the transformation formula has an importance of its own as a way to analyze the impact of a new unit on the performance of the electric power system. Such analysis is a major factor in various planning problems, such as reserve planning, capacity expansion decisions, evaluating alternative investment programs, and bid evaluation for power plants.

8.9 Using FORs versus Unconditional Outage Probabilities

Using the units FORs q_n, $n = 1, 2, \ldots, N$, results in a "potential" rather than an "actual" ELDC. The value of p_F added to the combined demand, in this case, is calculated under the assumption that all the units may be called to service at all times, and therefore represents the potential outage level rather than the actual one. The potential ELDC approaches in value the actual ELDC as the combined demand p_C increases. The two curves become identical for $p_C \geq p_{\text{ins}}$. Thus, for example, when calculating the LOLP or the LOEP of the system the potential ELDC yields the exact values, since at this level of demand all units are called for service. The reader will find that in all calculations of the following chapters, both those dealing with estimating reliability measures and those dealing with estimating costs, the FORs should be used if one desires to obtain accurate results.

The unconditional outage probabilities, q_n^*, $n = 1, 2, \ldots, N$, can be used when trying to derive an ELDC which is closer to the actual ELDC at off-peak combined demand levels. The resulting ELDC, however, will not be identical to the actual one at levels of peak demand since the outage probabilities of intermediate-load and peaking units is greatly underestimated by q_n^*. Base-load units, when not in maintenance, are required for service almost at all times and their q_n and q_n^* are practically identical. Thus their contribution to the ELDC is practically the same whether one uses q_n or q_n^*.

The probabilities q_n^* are dependent on the load factors of the units. Therefore, whenever the location of a unit in the merit order is unknown or uncertain, the individual load factor found as a result of the computation, using q_n^*, should be compared to the initial assumption. If the two are not satisfactorily close, the computation procedure should be redone with a more suitable assumption.

8.10 The Effect of Removing a Unit from the System

When analyzing the impact of possible improvements in an existing unit it is necessary first to remove the unit from the system in order later on to consider it as a "new" unit for the purpose of the analysis. In other words, we have to initiate the sensitivity-analysis procedure by calculating the ELDC corresponding to the system with the removed unit. Solving for $\bar{t}(p_C)$ in the transformation formula, the resulting ELDC is given by

$$\bar{t}(p_C) = \frac{\hat{t}(p_C) - - q\bar{t}(p_C - p_p)}{1 - q}. \qquad (8.19)$$

In this case, $\hat{t}(p_C)$, the ELDC prior to removing the unit from the system, is the known curve, and we are seeking $\bar{t}(p_C)$, the ELDC of the system after the unit has been removed. The calculations in this case are slightly more complicated than in the previous case since $\bar{t}(p_C - p_p)$, which appears in the right-hand side of the formula, is not known. However, $\bar{t}(p_C - p_p)$ can be obtained from the above formula by substituting $p_C - p_p$ for p_C, yielding

$$\bar{t}(p_C - p_p) = \frac{\hat{t}(p_C - p_p) - q\bar{t}(p_C - 2p_p)}{1 - q}, \qquad (8.20)$$

and one can continue deriving $\bar{t}(p_C - np_p)$, $n = 1, 2, \ldots$, until $p_C - np_p$ becomes zero or negative, at which point

$$\bar{t}(x) = 8760, \qquad x \text{ nonpositive}. \qquad (8.21)$$

Then by working backward, the new ELDC is obtained rather easily. If $q = 0$, we again obtain $\bar{t}(p_C) = \hat{t}(p_C)$.

Besides its role in sensitivity analysis, the procedure described above also has a merit of its own as a means of analyzing changes in performance measures when a unit is removed from the system. Such an analysis is required, for example, for unit retirement decisions.

Having described the basic transformation formulas, we will now expand on their application to the study of sensitivity of various measures of system performance to changes in the generation characteristics.

8.11 ELDC and Sensitivity Analysis

Sensitivity analysis refers to the task of estimating the impact of perturbing the data on the performance measures of the system. Such an analysis is particularly important in supporting planning decisions in electric power systems because of the high variability of data resulting from uncertainties about the future and the long planning horizon.

Being sensitive to changes in both demand and generation characteristics, the ELDC has proven to be most useful in providing sensitivity information in electric power systems. The analysis is carried out by studying the impact of perturbations on the system characteristics through their reflection in the ELDC. A host of such analyses, with respect to reliability measures, have been described by Vardi et al. (35).

In the following chapters methods for determining the change in reliability and cost performance of an electric power system resulting from perturbations in the generation characteristics will be developed and demonstrated. These will include the introduction of a new unit, the retirement of an old one, and changes in the size, FOR, or maintenance schedule of any unit in the system. The basic approach will be to consider each unit whose characteristics we want to investigate as a new unit, and generate the sensitivity information by successively adding units to the system, each time perturbing one or several of their parameters, and studying the resulting change in cost and reliability. More explicitly, the procedure contains the following steps:

1. Determine the ELDC resulting from augmenting the system by a unit with given parameters, and evaluate the new cost and reliability measures using the procedures described in this chapter.
2. Vary one or several of the parameters of the concerned unit, and repeat the analysis in (1), assuming all other parameters remain constant.
3. Compare the resulting performance measures to arrive at the required sensitivity information.

In the case that one is interested only in the sensitivity of the LOLP, there is no need to evaluate the whole ELDC for the augmented system, but rather the point on the new ELDC corresponding to the installed capacity of the system. It turns out that in such a case one can derive the sensitivity information in closed-form expressions.

8.12 ELDC versus Simulation

In many respects, the ELDC can be thought of as an analytical simulation model, serving as a substitute for digital simulation models currently used for detailed and depth analysis of electric power systems. The benefits of using an analytical rather than a simulation model are two: (a) it yields more insight into the system; and (b) it involves less calculational effort.

8.13 The Advantages of the ELDC

The ELDC has been successfully applied so far to calculating the expected cost according to merit order and system reliability (19). Recent works by the present authors and others have opened up additional potential applications of the ELDC (37, 40, 41). In summarizing the attributes of the ELDC the following are the reasons which render it such a versatile and powerful tool for electric power system analysis:

1. Forced outages and maintenance requirements are accounted for in an accurate manner which realistically reflects the impact of these factors on the performance measures of the system.
2. Better information is provided on the number of hours each unit is scheduled for generation in any given period, resulting in improved estimates of the expected energy delivered by each unit in the system.
3. The provision to include random deviations of demand in the ELDC contributes to increase further the accuracy of the estimated system's performance measures, especially for short-range (1–3 years) planning horizons, where precise prediction of these measures is necessary.
4. By its definition, the ELDC is a function both of the demand and the generation characteristics of the system, the former through the deterministic and random components of customers' demand, the latter through the requirements for maintenance and forced outages. Consequently, the ELDC is sensitive to perturbations of all important parameters of the system.
5. Last, but not least, the ELDC is very easy to calculate and manipulate. Furthermore, the ELDC can be obtained to varying degrees of accuracy, depending upon the type of application involved (35). Even in its approximate form, the ELDC provides better information than its predecessor, the LDC. In most cases, an approximate ELDC can be used to screen out relevant possibilities, which can then be further analyzed using an exact ELDC.

9

Reliability Analysis by the Equivalent Load–Duration Curve

9.1 Introduction

In this chapter an analytical rather than simulational approach is presented for performing reliability-related sensitivity and trade-off analyses of investment policies of electric power systems. The method of analysis is based on the ELDC. Since by its definition the ELDC is a function of both customers' demand and generation characteristics of the power system, changes in the characteristics of the system affect the ELDC. This property of the ELDC is exploited to derive a host of reliability-related sensitivity analyses by studying the impact of perturbing the characteristics of the system through their reflection in the ELDC. The application of these analyses to the problem of determining reserve requirements is also illustrated.

9.2 Computing the Basic Reliability Measures through the ELDC

The use of the ELDC to calculate the LOLP and the LOEP has been discussed in various publications (3). Since by definition any combined demand beyond the installed capacity is an unsupplied load, we have

$$\text{LOLP} = \bar{t}(p_{\text{ins}}) \tag{9.1}$$

where

p_{ins} = the installed capacity of the system.

By the same logic, the complementary cumulative density function of unserved energy (known in short as the *nonsupply distribution*) is simply the tail of the ELDC above the installed capacity level. If $S(x)$ denotes the nonsupply distribution (the number of hours in a given period during which the shortage equals or exceeds x MW), we have

$$S(x) = \bar{T}(x + p_{\text{ins}}), \qquad x \geq 0. \tag{9.2}$$

By definition, the area under the nonsupply distribution equals the expected amount of unserved energy (known as the LOEP), i.e.,

$$\text{LOEP} = \int_{x=0}^{p_{\text{Dmax}}} S(x)\,dx, \tag{9.3}$$

where p_{Dmax} is the peak demand for power.

The capacity margin Δ at any given point in time has been defined as the difference between available capacity and customers' demand. A capacity margin distribution is given graphically in figure 4.6.

In general,

$$\Delta = p_{\text{ins}} - p_{\text{M}} - p_{\text{F}} - (p_{\text{D}} + p_{\text{R}}) \tag{9.4}$$

or, equivalently,

$$\Delta = p_{\text{ins}} - p_{\text{C}}, \tag{9.5}$$

where $p_{\text{C}} = p_{\text{D}} + p_{\text{R}} + p_{\text{M}} + p_{\text{F}}$ is the combined demand for power.

By the definition of the ELDC it is apparent that

$$\tilde{t}(\Delta) = \bar{T}(p_{\text{ins}} - \Delta), \tag{9.6}$$

where $\tilde{t}(\Delta)$ is the capacity margin distribution, interpreted as the number of hours, in a given period, during which the capacity margin equals or exceeds Δ MW.

9.3 Application of the ELDC to Sensitivity Analysis

Determining the change in reliability performance resulting from changes in any of the parameters is important for decision making. This process is referred to in this section as *sensitivity analysis.*

The Impact of an Added Unit on the System's Reliability

When adding a new unit to the system, the new LOLP is computed as follows. Let

p_{insN} = the new installed capacity (i.e., after adding the unit to the system),

p_{insO} = the previous installed capacity,

p_p = the capacity of the added unit,

q = outage probability of the added unit.

Then by definition

New LOLP = $\hat{t}(p_{\text{insN}})$,

Old LOLP = $\bar{t}(p_{\text{insO}})$.

Using the transformation formula, we have

$$\hat{t}> (p_{\text{insN}}) = (1-q)\,\bar{t}(p_{\text{insN}}) + q\bar{t}(p_{\text{insN}} - p_p) \tag{9.7}$$

or

$$\text{New LOLP} = (1 - q)\,\bar{t}(p_{\text{ins N}}) + q\,(\text{Old LOLP}). \tag{9.8}$$

This formula provides us with a very easy way to determine the reliability of a power system measured by its LOLP as the configuration of the system is changed. An important use of this formula for determining reserve requirements is discussed in the next section.

Figure 9.1 demonstrates the changes in the ELDC and LOLP when adding a unit.

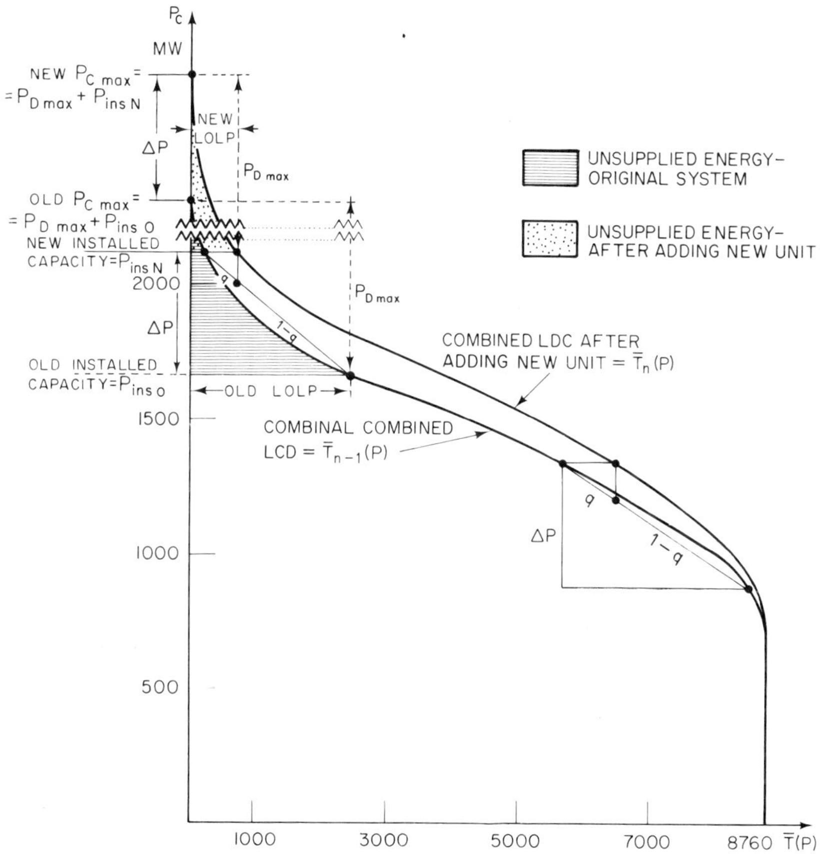

Figure 9.1 The effect of adding a unit on the LOLP.

The Effect of Unit Size on Reliability—An Example

The sensitivity of the reliability measures to changes in unit size is a common problem in the analysis of electric power systems. To illustrate this we assume the following two possibilities for augmenting the system:

1. two units, each with capacity p_p MW and outage probability q;
2. one unit with double the capacity ($2p_p$), and the same outage probability.

The problem is to determine the reliability level, measured by the LOLP, of the two alternatives, while ignoring differences in maintenance flexbility of the additional units.

Let $t_1(p)$ and $t_2(p)$ denote the new ELDC in cases (1) and (2), respectively.

Using the transformation formula and adding the first unit with capacity p_p MW, we have

$$\hat{t}(p_{\text{ins}} + p_p) = (1 - q)\bar{t}(p_{\text{ins}} + p_p) + q\bar{t}(p_{\text{ins}}). \tag{9.9}$$

Adding the second unit with capacity p_p MW, we have

$$t_1(p_{\text{ins}} + 2p_p) = (1 - q)\hat{t}(p_{\text{ins}} + 2p_p) + q\hat{t}(p_{\text{ins}} + p_p). \tag{9.10}$$

Substituting and combining terms yields

$$t_1(p_{\text{ins}} + 2p_p) = q^2\bar{t}(p_{\text{ins}}) + 2(1 - q)q\bar{t}(p_{\text{ins}} + p_p) + (1 - q)^2\bar{t}(p_{\text{ins}} + 2p_p). \tag{9.11}$$

Shifting now to case (2) and adding one unit with double capacity, we have from the transformation formula

$$t_2(p_{\text{ins}} + 2p_p) = (1 - q)\bar{t}(p_{\text{ins}} + 2p_p) + q\bar{t}(p_{\text{ins}}). \tag{9.12}$$

Taking the difference between the LOLPs obtained in the two cases, we have

$$\text{LOLP}_2 - \text{LOLP}_1 = t_2(p_{\text{ins}} + 2p_p) - t_1(p_{\text{ins}} + 2p_p). \tag{9.13}$$

Therefore, the difference between the LOLP of alternatives (1) and (2), ignoring maintenance flexibility of the additional units, is given by

$$\text{LOLP}_2 - \text{LOLP}_1 = (1 - q)q[\bar{t}(p_{\text{ins}} + 2p_p) - 2\bar{t}(p_{\text{ins}} + p_p) + \bar{t}(p_{\text{ins}})]. \tag{9.14}$$

Since the tail of the ELDC is a convex function, the quantity in the square brackets is always positive, and thus the LOLP, when adding one unit to the system with capacity $2p_p$, is larger than the LOLP obtained when adding two units, each with half the capacity. Even though this is not a surprising result, the advantage of the above method is that it tells us by how much the reliability is affected.

When precise results are required, there is a need to account for maintenance requirements of the additional unit in an accurate manner by recalculating the ELDC. This can be done following the procedure outlined in Chapter 8.

By carrying out the calculations for various values of p_p, a trade-off curve between reliability and unit size can be obtained for a given level of outage probability.

Effect of Changing Maintenance Schedule

Alternative maintenance schedules will most likely result in different reliability levels. The impact of changing the maintenance schedule on the reliability of the system can be analyzed using the ELDC. The transformation formula in this case is used for finding the entire ELDC for each maintenance schedule. While involving numerous computations, this analysis enables the decision maker to examine alternative maintenance schedules for the purpose of finding the one which yields the best level of reliability.

Sensitivity to Changes in Demand Characteristics

The demand for electric power is another major factor which influences the reliability level of the system. Changes in demand, such as a sudden increase or decrease in peak demand, change of the load factor, change of the load shape, etc., might affect the system reliability in a way no less pronounced than changes in generation characteristics. Being also a function of the demand characteristics, the ELDC provides us with an effective tool for analyzing the impact of such changes on the reliability. The ELDC must be regenerated for each change in demand characteristics. By using the recursive equations, the task of deriving the new ELDC is greatly facilitated, as the only expression affected by the changes in demand is $\bar{t}_0(p_C)$, a partial ELDC containing the deterministic and random components of demand as well as maintenance requirements. $\bar{t}_0(p_C)$ is easily obtained since it involves an evaluation of only one integral for each level of p_C.

Finding the Reserve Required to Meet a Given Level of LOLP

This problem will be answered for the case for which peaking units are added to the system to provide for the reserve versus the case where base and intermediate units are installed to increase the reserve margin.

For those peaking units which are working only a small fraction of their time, one can approximate their impact on the ELDC by assuming that their outage probabilities equal zero. Substituting $q = 0$ in the transformation formula, we obtain

$$\hat{t}(p_C) = \bar{t}(p_C), \tag{9.15}$$

i.e., the new ELDC equals the previous ELDC; this is an approximate result since peaking units do not contribute significantly to demand for maintenance and forced outages. If peaking units are used to provide the reserve capacity, the amount of generation capacity required to meet a given LOLP can be read directly off the ELDC. In other words, for a given LOLP we solve for the new installed capacity $p_{\text{ins N}}$ satisfying

$$\text{LOLP} = \bar{t}(p_{\text{ins N}}). \tag{9.16}$$

The peak demand is then subtracted from the required installed capacity to yield the necessary level of reserve. This is a rather simple procedure, based on a quick-and-dirty approximation.

If base-load or intermediate-load units are installed to increase the reserve margin a searching procedure is necessary for finding the required reserve level since these units contribute significantly to the demand for forced outages and maintenance, thus changing the ELDC. For each unit added or removed from the system, the new LOLP is derived using the procedure described in chapter 8. If the required LOLP is met, the search is terminated; otherwise, a new configuration of units is examined, and the process is repeated until the design criterion is met.

Finding the Reserve Required for Meeting a Given Level of Unsupplied Energy

Rather than meeting a given level of LOLP, we might want to find the reserve required to meet a given level of expected unsupplied energy. The general procedure would be to increase the reserve in the system until the area under the tail of the ELDC, above the installed capacity, equals the required level of unsupply. If base-load or intermediate-load units are added to increase the reserve margin, some searching procedure is required for finding the necessary level of reserve. The transformation formula can be very helpful in reducing the number of calculations required to support such a decision. If, however, peaking units are used as a means of providing for the reserve, the analysis becomes quite simple, provided the following two assumptions hold:

- the peaking units are working only a small fraction of the time during the period and we assume that their outage probabilities equal zero;
- the tail of the ELDC above the installed capacity is exponential.

Denoting by S_0 (in MW h) the required level of expected unsupplied energy, and by S (in MW h) the expected unsupplied energy of the present system, the level of LOLP which yields an expected unsupplied energy of S_0 under the above assumptions is given by

$$\text{New LOLP} = \text{Old LOLP}\,\frac{S_0}{S} \tag{9.17}$$

since the areas under the exponential curve are proportional to the ordinates.

Having obtained the LOLP required to maintain a given level of S_0, the required installed capacity p_{insN} is determined by solving

$$\text{New LOLP} = \bar{t}(p_{\text{insN}}). \tag{9.18}$$

10

Operating Cost Analysis by ELDC

Several approaches to calculating the expected operating cost of an electric power system were mentioned and briefly described in chapter 5. The probabilistic simulation approach is one of the most sophisticated methods and is based on the ELDC (41). This method, as well as a refinement of it which employs the incremental, rather than the merit order, loading, is presented in this chapter.

10.1 Operating Cost Calculation of a System with Supply Uncertainty under Merit Order Loading

The probabilistic simulation method (19) is probably the first analytical model which accurately accounts for forced outages in calculating the operating cost of the system under the merit order loading. This procedure is reviewed in the following.

Figure 3.4 describes an LDC of a given system along with the merit order loading, with unit 1 being the most efficient unit, unit 2 being the next most efficient unit, and so on. Letting

$t(p_L)$ = the LDC evaluated at point p_L, interpreted as the fraction of time, in a given period, during which customers' demand equals or exceeds p_L,

E_i = the energy delivered by unit i (numbered according to the merit order),

p_i = the capacity of unit i,

q_i = the outage probability of unit i,

and ignoring maintenance requirements, the expected energy delivered by unit 1 is given by

$$E_1 = (1 - q_1) \int_0^{p_1} t(p_L)\, dp_L. \tag{10.1}$$

In calculating the energy delivered by the second unit in the merit order, two components of costs are to be considered. When unit 1 is available, with a probability of $(1 - q_1)$, unit 2 will be loaded into production between the loads p_1 and $p_1 + p_2$; however, when unit 1 is out of service, because of forced outage, with a probability q_1, unit 2 will occupy the first position in the merit order and will then be loaded for generation between the loads 0 and p_2, as described in figure 3.5. The expected energy delivered by unit 2 is therefore

$$E_2 = (1 - q_2)\left[(1 - q_1) \int_{p_1}^{p_1+p_2} t(p_L) + q_1 \int_{0}^{p_2} t(p_L)\, dp_L\right]. \tag{10.2}$$

Substituting

$$\int_{0}^{p_2} t(p_L)\, dp_L = \int_{p_1}^{p_1+p_2} t(p_L - p_1)\, dp_L, \tag{10.3}$$

we have

$$E_2 = (1 - q_2)\left\{\int_{p_1}^{p_1+p_2} [(1 - q_1)\, t(p_L) + q_1 t(p_L - p_1)]\, dp_L\right\}. \tag{10.4}$$

It has been shown (see equation (8.17)) that the expression in brackets under the integral sign is a partial equivalent load–duration curve (ELDC) containing customers' demand and requirements for forced outages of the first unit. Following our notation, we denote this expression by $\bar{t}_1(p_L)$ and obtain

$$E_2 = (1 - q_2) \int_{p_1}^{p_1+p_2} \bar{t}_1(p_L)\, dp_L. \tag{10.5}$$

Continuing in this fashion, it can be shown that

$$E_n = (1 - q_n) \int_{\sum_{i=1}^{n-1} p}^{\sum_{i=1}^{n} p} \bar{t}_{n-1}(p_L)\, dp_L, \tag{10.6}$$

where $\bar{t}_n(p_L)$ is the ELDC containing customers' demand and the demand for forced outages of the first n units.

The expected energy delivered by each unit in the system under merit order loading is therefore closely related to the ELDC. By using the recursive formula from the previous chapter we get

$$\bar{t}_n(p_L) = (1 - q_n)\,\bar{t}_{n-1}(p_L) + q_n \bar{t}_{n-1}(p_L - p_n). \tag{10.7}$$

The partial ELDCs involved in calculating the expected delivered energies, and hence the operating costs, can be obtained in a very efficient manner using this recursive formula.

The total operating cost of the system is derived by multiplying the expected energy delivered by each unit by the corresponding average production cost and adding up the products for all units.

If maintenance requirements are to be taken into consideration, the period involved is to be partitioned into subperiods of constant maintenance. The operating costs are then calculated for each subperiod separately, and combined to yield the total operating cost for the whole period. Provisions for incorporating hydroelectric stations and pumped-storage operations into the calculation have also been developed (15, 19).

10.2 Operating Cost Calculation under Incremental Loading for a System Subject to Forced Outages

The approach presented in chapter 5, dealing with operating cost calculation for a system under incremental cost loading without forced outages, will now be extended to a system which is subject to maintenance and forced outages. One possible way for calculating the operating cost of such a system, under incremental loading procedure, is to construct first the marginal cost functions for each combination of up and down (due to forced outages) units. Given these cost functions, one can calculate the operating cost for each of the combinations, using the previously described method, and then multiply each cost by the probability of its occurrence and sum the results to obtain the expected total operating cost of the system for the given period. This enumeration method is a lengthy one, however, involving 2^N derivations of marginal cost functions, equal to the number of up and down combinations in an N-unit system, which increases exponentially with the number of units in the system. Figure 10.1 shows a system of two units with forced outage rates q_1 and q_2 resulting in four different combinations of units up and down. The figure presents the three different marginal cost functions (in the fourth combination, where the two units are down, no energy is delivered), and their associated probabilities. For every given load p the expected total marginal cost is obtained by multiplying the probability of each of the three functions by its value at p, and then summing.

A more efficient method, resulting in substantial savings in the amount of calculations required to derive operating cost figures for a system working under the incremental loading procedure, will now be derived.

Basic Notation

We first rearrange the units in the system in an ascending order of their minimal marginal cost, so that:

$$\min \{c'_{i+1}\} \geq \min \{c'_i\}, \qquad i = 1, 2, \ldots, N, \tag{10.8}$$

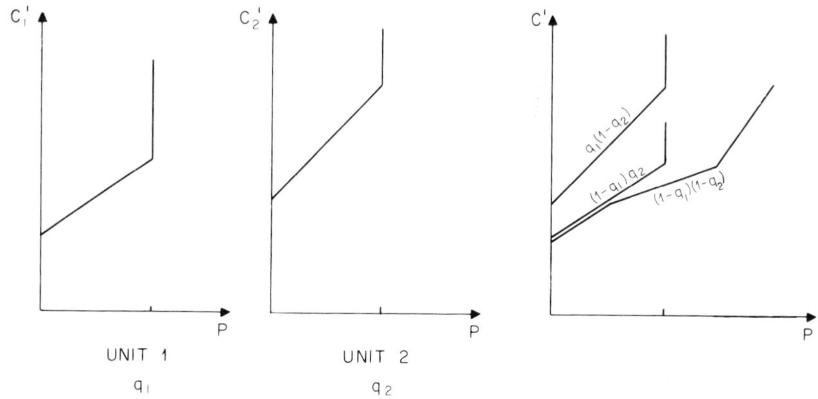

Figure 10.1 Finding the equivalent marginal cost of the system (example).

where c_i' denotes the marginal cost function of generator i in this new order. The minimal marginal cost for generator i is obtained by evaluating the marginal cost function at the point of minimum output, assuming that c_i' is an increasing function.

For each unit i let Q_i denote the subset of generators, with index $j > i$, having at least one common marginal cost value with generator i. Under the incremental loading procedure these units may be loaded into production with generator i in at least some cases in which generator i is not loaded to full capacity. Figure 10.2 demonstrates the definition of the subsets Q_i, $i = 1, 2, \ldots, N$. For any given subset Q_i containing n_i generators, there exists

$$\max\{c_i'\} \geq \min\{c_j'\}, \qquad j = i+1, \ldots, \qquad i + n_i \leq N. \tag{10.9}$$

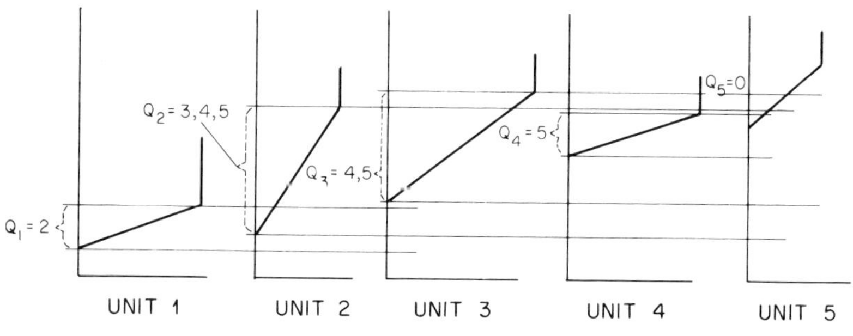

Figure 10.2 Example of the definition of Q_i ($N = 5$).

For most electric power systems, which are composed of many units with varying characteristics, the number of units in each subset Q_i is smaller than the number of generators in the system. Denoting by n_i the number of units in subset Q_i, the number of possible up and down (due to forced outages) combinations of these units is 2^{n_i}. Since $n_i < N$, $2^{n_i} \ll 2^N$. Let R_i, $i = 1, 2, \ldots, N$, be the set of all combinations of up and down units of subset Q_i, with r_i serving to denote one element of the subset having an occurrence probability p_{r_i}. The marginal cost function for each combination $r_i \in R_i$ is defined as follows.

Let $c'_{r_i}(x)$ denote the marginal-cost function resulting from loading unit i together with the operative generators (units) of combination r_i to a given generation level x. Note that by definition unit i is not a member of the combination $r_i \in R_i$; however, when calculating $c'_{r_i}(x)$, unit i is considered as a member of all combinations $r_i \in R_i$ (e.g., suppose r_j is the combination for which all members of Q_i are down; then $c'_{r_j}(x) = c'_i(x)$). The marginal cost functions, one for each combination in R_i, are obtained by adding the abscissas of the marginal cost functions corresponding to the working units in each combination. There are 2^{n_i} marginal cost functions for each subset.

Calculating the Expected Total Operating Cost of the System

Given the above definitions, a phased approach to calculating the expected operating cost of the system will be devised.

When unit 1 is working, with probability $(1 - q_1)$, then for any given combination $r_1 \in R_1$ (recall that by the definition of Q_1, unit 1 is *not* a member of the combinations in R_1) the incremental operating cost for the system is

$$c'_{r_1}(p_L)\, t(p_L)\, dp_L, \qquad p_L \le p_1. \tag{10.10}$$

The contribution to the expected total operating cost of the units in the subset Q_1 which are loaded to generation along with unit 1 in the region $0 \le p_L \le p_1$ is given by

$$\mathrm{TC}_1 = (1 - q_1) \int_0^{p_1} \left[\sum_{r_1 \in R_1} p_{r_1} c'_{r_1}(p_L) \right] t(p_L)\, dp_L. \tag{10.11}$$

The expression in the square brackets is clearly the expected marginal cost of meeting the demand by the units in subset Q_1 together with unit 1. Denoting this expression by $\bar{c}'_{R_1}(p_L)$, we have

$$\begin{aligned} \mathrm{TC}_1 &= (1 - q_1) \int_0^{p_1} \bar{c}'_{R_1}(p_L)\, t(p_L)\, dp_L \\ &= (1 - q_1) \int_0^{p_1} \bar{h}_0(\bar{c}'_{R_1}(p_L))\, dp_L, \end{aligned} \tag{10.12}$$

where $\bar{h}_0(\bar{c}'_{R_1}(p_L)) = \bar{c}'_{R_1}(p_L)\,t(p_L)$ by definition.

Moving on to the second unit, we can identify three components in calculating the contribution to the expected total operating cost when unit 2 is available. The first component refers to the case for which the first unit is working and we are considering the region $0 \le p_L \le p_1$. This component has already been taken care of in the calculation of TC_1. The second component refers to the case for which the first unit is available and unit 2 along with the units of subset Q_2 are loaded for generation in the region $p_1 \le p_L \le p_1 + p_2$. The third component refers to the case for which unit 1 is out of service because of forced outage and we are considering the region $0 \le p_L \le p_2$. Denoting the contribution of the latter two components to the expected operating cost by TC_2, we have

$$TC_2 = (1 - q_2)\Bigg[(1 - q_1) \int_{p_1}^{p_1+p_2} \sum_{r_2 \in R_2} P_{r_2} c'_{r_2}(p_L)\,t(p_L)\,dp_L + q_1 \int_0^{p_2} \sum_{r_2 \in R_2} P_{r_2} c'_{r_2}(p_L)\,t(p_L)\,dp_L\Bigg]. \quad (10.13)$$

Substituting $\bar{c}'_{R_2}(p_L)$ for the expected marginal cost of meeting demand by the units in subset Q_2, rearranging terms, and transforming variables, we obtain

$$TC_2 = (1 - q) \int_{p_1}^{p_1+p_2} [(1 - q_1)\,\bar{c}'_{R_2}(p_L)\,t(p_L) + q_1\,\bar{c}'_{R_2}(p_L - p_1)\,t(p_L - p_1)]\,dp_L \quad (10.14)$$

and, by the definition of the function $\bar{h}_0$,

$$TC_2 = (1 - q_2) \int_{p_1}^{p_1+p_2} [(1 - q_1)\,\bar{h}_0(\bar{c}'_{R_2}(p_L)) + q_1 \bar{h}_0(\bar{c}'_{R_2}(p_L - p_1))]\,dp_L. \quad (10.15)$$

Denoting the expression in the square brackets by $\bar{h}_1(\bar{c}'_{R_2}(p_L))$, we have

$$TC_2 = (1 - q_2) \int_{p_1}^{p_1+p_2} \bar{h}_1(\bar{c}'_{R_2}(p_L))\,dp_L. \quad (10.16)$$

Continuing in this fashion we can obtain the general form

$$TC_n = (1 - q_n) \int_{\sum_{i=1}^{n-1} p_i}^{\sum_{i=1}^{n} p_i} \bar{h}_{n-1}(\bar{c}'_{R_n}(p_L))\,dp_L, \quad (10.17)$$

where

$$\bar{h}_n(\bar{c}'_{R_k}(p_L)) = (1 - q_n)\bar{h}_{n-1}(\bar{c}'_{R_k}(p_L)) + q_n\bar{h}_{n-1}(\bar{c}'_{R_k}(p_L - p_n)),$$
$$n = 1, 2, \ldots, N - 1. \tag{10.18}$$

TC_n is the contribution to the expected operating cost of the system by the units in subset Q_n which have not been taken care of in previous calculations. The expected total operating cost of the system is obtained by adding all the contributions, i.e.,

$$TC = \sum_{i=1}^{N} TC_i. \tag{10.19}$$

Computationally, in order to derive TC_n we first have to generate the sequence of functions

$$\bar{h}_i(\bar{c}'_{R_{i+1}}(p_L)), \qquad i = 0, 1, 2, \ldots, n - 1. \tag{10.20}$$

The bulk of the computational effort is in deriving the N expected marginal cost curves $\bar{c}'_{R_i}(p_L)$, each of which involves 2^{n_i} marginal-cost functions. However, recall that $n_i \ll N$, 2^{n_i} is a relatively small number, much smaller than the 2^N elements involved in the direct enumeration method. As an example, consider a system with $N = 20$ units for which the generating units can be divided into 20 subsets with 5 units in each subset on the average. Then $2^5 = 32$ combinations will have to be considered to calculate the expected marginal cost function for each subset Q_i, or 640 elements in all versus the 2^{20} combinations involved in the enumeration method. A substantial saving indeed. Once the expected marginal cost functions have been derived for all subsets, the task of calculating TC_n becomes rather simple. A computer can be programmed to carry out the calculations quite efficiently.

A further simplification of the calculation process is available if, instead of using the expected marginal cost functions $\bar{c}'_{R_n}(p_L)$, we consider their average values, given by

$$\bar{c}'_n = \int_0^{\sum_{i=1}^{n} p_i} \bar{c}'_{R_n}(p_L) f(p_L)\, dp_L, \tag{10.21}$$

where $f(p_L)$ is the load density function of customers' demand, obtained very easily from the LDC (see equation (3.2)) (7).

$\bar{c}'_n$ can be interpreted as representing the average expected marginal cost of meeting demand by the units in subset Q_n together with unit n. By plugging $\bar{c}'_n$ into equation (10.17) instead of $\bar{c}'_{R_n}(p_L)$, it can be factored out, reducing the expression for TC_n to

$$TC_n = (1 - q_n)\bar{c}'_n \int_{\sum_{i=1}^{n-1} p}^{\sum_{i=1}^{n} p} \bar{t}_{n-1}(p_L)\, dp_L, \tag{10.22}$$

where $\bar{t}_n(p_L)$ is the partial ELDC containing customers' demand (both deterministic and random components) and the demand for outages of the first n units. A recursive equation for calculating $\bar{t}_n(p_L)$ has been presented in the previous chapter.

The calculation in this case, for an N-unit system, involves the derivation of N average expected marginal costs, one for each subset Q_n, calculation of N partial ELDCs, and N integrations. This is a rather simplified procedure.

We note the similarity between the expression for TC_n in equation (10.22) and the corresponding expression obtained when merit order loading is used. It should be emphasized, however, that even this last approximation exhibits an improvement in calculating cost figures compared with merit order loading, as the generating units are classified into groups according to their marginal cost functions and a different average marginal cost is used for each group. When exact results are required, the calculation must be carried out using the functions $\bar{c}'_{R_n}(p_L)$, at the expense of increased computational effort.

Finding the Expected Fuel Cost for Each Unit in the System

The expected fuel cost for each unit in the system can be obtained by modifying the calculation of TC_n to yield the contribution of the unit to the operating cost, and then adding the contributions for all relevant subsets.

Based on the argument presented in the previous chapter, the contribution of unit k to the operating cost of the system, for a given combination $r_n \in R_n$ and a given load p_L, in the interval

$$\sum_{i=1}^{n-1} p_i \le p_L \le \sum_{i=1}^{n} p_i, \tag{10.23}$$

is

$$c'_{r_n}(p_L)\frac{dp_{L_k}(r_n)}{dp_L}\,t(p_L)\,dp_L, \tag{10.24}$$

where $P_{L_k}(r_n)$ is the load of unit k and $dp_{L_k}(r_n)/dp_L$ is the ratio between the reciprocal slope of the marginal cost function of unit k at demand level p_L and the reciprocal slope of the marginal cost function of all working units in combination $r_n \in R_n$ at the given demand level (see section 5.4). Since the composition of operative and nonoperative units in each combination is different, the above ratio is a function of r_n and should be recomputed for each combination. For those combinations for which unit k is down the above ratio is zero.

The expected contribution of unit k, $k \in Q_n$, to the operating cost for the whole load interval is, therefore,

$$\int_{\sum_{i=1}^{n-1} p_i}^{\sum_{i=1}^{n} p_i} \left[\sum_{r_n \in R_n} p_{r_n} c'_{r_n}(p_L) \frac{dp_{L_k}(r_n)}{dp_L} \right] t(p_L)\, dp_L. \tag{10.25}$$

The expression in the square brackets can be interpreted as the kth unit, $k \in Q_n$, expected marginal cost of meeting the demand in the interval $\sum_{i=1}^{n-1} p_i \leq p_L \leq \sum_{i=1}^{n} p_i$.

Denoting this expression by $\bar{c}'_{R_n}(p_L, k)$, the above expression becomes

$$\int_{\sum_{i=1}^{n-1} p_i}^{\sum_{i=1}^{n} p_i} \bar{c}'_{R_n}(p_L, k)\, t(p_L)\, dp_L. \tag{10.26}$$

To obtain the total expected operating cost of unit k, it is necessary to sum its contributions to the satisfied demand over all intervals in which the unit is loaded for generation. This is done by using the procedure described for calculating the operating costs of the system, except that the computations are carried out separately for each unit k and $\bar{c}'_{R_n}(p_L, k)$ is substituted for $\bar{c}'_{R_n}(p_L)$. Denoting the contribution of unit k to the expected operating costs incurred by the units in subset Q_n, $k \in Q_n$, which have not been accounted for in previous calculations by $\bar{c}^n_k$, we have

$$\bar{C}^n_k = (1 - q_n) \int_{\sum_{i=1}^{n-1} p_i}^{\sum_{i=1}^{n} p_i} \bar{h}_{n-1}(\bar{c}'_{R_n}(p_L, k))\, dp_L, \tag{10.27}$$

where $\bar{h}_n\,(\bar{c}'_{R_n}(p_L, k))$ is computed recursively using the formula

$$\begin{aligned} \bar{h}_n(\bar{c}'_{R_j}(p_L, k)) &= (1 - q_n)\, \bar{h}_{n-1}(\bar{c}'_{R_n}(p_L, k)) \\ &\quad + q_n \bar{h}_{n-1}(\bar{c}'_{R_j}(p_L - p_n, k)), \\ \bar{h}_0(\bar{c}'_{R_1}(p_L, k)) &= \bar{c}'_{R_1}(p_L, k)\, t(p_L). \end{aligned} \tag{10.28}$$

The expected operating cost of unit k is therefore

$$\bar{C}_k = \sum_{n \in A} \bar{C}^n_k,$$

where A is the set of elements $k, k - 1, \ldots, k - j$, and j is the first unit satisfying $\max\{c'_{j-1}\} < \min\{c'_k\}$.

The procedure results in a relatively large volume of computations. If less accuracy is permitted, the amount of calculations can be reduced in a substantial manner by approximating the functions $\bar{c}'_{R_n}(p_L, k)$ by their expected values:

$$\bar{c}'_n(k) = \int_0^{\sum_{i=1}^{n} p} \bar{c}'_{R_n}(p_L, k) f(p_L)\, dp_L, \tag{10.29}$$

where $f(p_{\rm L})$ is the load density of demand. Substituting $\bar{c}'_n(k)$ for $\bar{c}'_{R_n}(p_{\rm L}, k)$ in the above formulas, the expression for expected operating cost of unit k reduces to

$$\bar{C}_k = \sum_{n \in A} (1 - q_n)\bar{c}'_n(k) \int_{\sum_{i=1}^{n-1} p_i}^{\sum_{i=1}^{n} p_i} \bar{t}_{n-1}(p_{\rm L})\, dp_{\rm L}, \tag{10.30}$$

where $\bar{t}_n(p_{\rm L})$ is the partial ELDC containing customers' demand (both deterministic and random components) and the demand for forced outages of the first n units.

Given the unit cost of fuel for each generator, the expected amount of fuel consumed by the unit can be easily determined.

Calculating the Expected Energy Delivered by Each Unit in the System

The expected energy delivered by each unit in the system can be obtained as a special case of the cost calculations by substituting $c'_{r_n}(p_{\rm L}) = 1$ for all levels of $p_{\rm L}$. The incremental energy attributed to unit k, $k \in Q_n$, for a given combination $r_n \in R_n$ and a given load in the interval

$$\sum_{i=1}^{n-1} p_i \le p_{\rm L} \le \sum_{i=1}^{n} p_i,$$

is therefore

$$\frac{dp_{{\rm L}_k}(r_n)}{dp_{\rm L}} t(p_{\rm L})\, dp_{\rm L}, \tag{10.31}$$

from which the expected contribution to the energy delivered by unit k, $k \in Q_n$, in that interval is given by

$$\int_{\sum_{i=1}^{n-1} p_i}^{\sum_{i=1}^{n} p_i} \left[\sum_{r_n \in R_n} p_{r_n} \frac{dp_{{\rm L}_k}(r_n)}{dp_{\rm L}} \right] t(p_{\rm L})\, dp_{\rm L}. \tag{10.32}$$

Denoting the expression in the square brackets by $\bar{m}_{R_n}(p_{\rm L}, k)$, and following the same procedure as before, we obtain

$$\bar{E}_k^n = (1 - q_n) \int_{\sum_{i=1}^{n-1} p_i}^{\sum_{i=1}^{n} p_i} \bar{h}_{n-1}(\bar{m}_{R_n}(p_{\rm L}, k))\, dp_{\rm L}, \tag{10.33}$$

where $\bar{E}_k^n$ is the expected contribution of unit k to the energy delivered by the units in subset Q_n, $k \in Q_n$, which have not been accounted for in previous calculations, and where $\bar{h}_{n-1}(\bar{m}_{R_n}(p_{\rm L}, k))$ is obtained using recursive equations similiar to equation (10.28).

The expected energy delivered by unit k is

$$\bar{E}_k = \sum_{n \in A} \bar{E}_k^n. \tag{10.34}$$

The approximate procedure for calculating $\bar{E}_k$, resulting in a substantial reduction in the computational volume, can also be applied in a similar fashion to the one suggested by equations (10.29) and (10.30).

Unit Commitment

System production costs are very sensitive to variations in unit commitment. Marsh et al. (23) have suggested a procedure to account for unit commitment in cost calculations. According to their approach, each period of constant maintenance should be further divided into time zones of constant commitment by grouping hours with identical commitments and arranging the load in each grouping into a load–duration curve. Then the operating cost calculations are carried out separately for each zone of constant commitment and constant maintenance using the above procedures, and summed up to yield the operating cost for the whole period. While providing us with more accurate cost measures, the method obviously results in an increased number of calculations.

11

Peaking Units Analysis by ELDC

11.1 The Nature of Peak Demand

The deterministic methods based on the LDC usually fail to describe the system performance at peak hours as they do not expect that peaking units, which are more expensive to operate, will ever be assigned to generation. In actual life, peaking units do operate a nonnegligible amount of time. Unit 6 in figure 11.1 is not loaded at all according to the system's load–duration curve, while in real life it will work at times when other units will be forced out. The ELDC provides much better information than the LDC on the expected fraction of time, or the expected number of hours in a given period, that individual units, including peaking units, are scheduled to operate.

Using the notion of the combined demand, we redefine the peak demand for electric power to include customers' load (both deterministic and random components) as well as requirements for maintenance and forced outages. With this definition, peaking conditions are not created by customers' demand alone, but also by the load imposed on the system by forced outages and maintenance. This is a more accurate way to define the peak period, for, as the customer and peaking units are concerned, if the electricity in demand is not supplied due to a shortage, it makes no difference whether this is because all the customers have suddenly increased their consumption, thus causing a rise in demand, or because a generator has suddenly gone down and was forced out of service, thus reducing the amount of capacity available for service. In both cases operative peaking units will be working. In fact, it can be shown (14) that peaking conditions may arise at any hour of the day, even at periods when customers' demand is low, because there is always a possibility, however small, of simultaneous outages of units (either because of preventive maintenance or breakdowns) that will bring the available capacity in the system to a level where it cannot accommodate all the demand for power. With this redefinition of peak demand, we realize that shortage situations will arise whenever the combined demand for power exceeds the installed capacity of the system. In all such situations the operative peaking units will be activated.

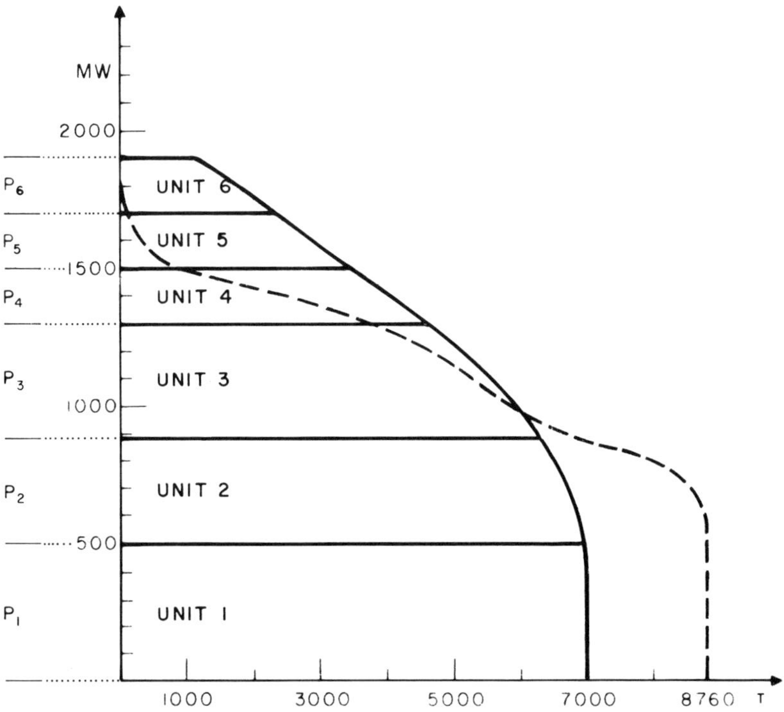

Figure 11.1 Customers' load–duration curve (LDC) versus plant loading–duration curve. Key: ——— units loading–duration curve; – – –, customers' load–duration curve.

The capability of the ELDC to describe accurately the system at peak periods is fundamental to sensitivity analysis since performance measures for electric power systems are often more sensitive to the shape of the ELDC during peak hours than during off-peak hours.

11.2 Applying the ELDC to Analyzing Peaking Units

The use of the ELDC instead of the LDC in the solution of the optimal mix problem is not feasible. This is because the ELDC is a function of the number and size of the units in the system, which are unknowns in the optimal mix problem. However, the ELDC can be used quite satisfactorily for approximate analysis of peaking units employing an optimal mix-type model.

For this purpose we assume that the outage probabilities of the peaking units approach zero.

To initiate the optimal mix procedure we first derive the distribution of the "remaining system load" to be satisfied by peaking facilities, i.e., the

LDC for the peak period which results after discounting the energy provided by base-load and intermediate-load units and after allowing for maintenance and forced outages requirements of these units.

We recall that we have already found the ELDC for the last peaking unit in the system (see equation (9.15)). If N denotes the position of this unit in the merit order, the function $\bar{t}_{N-1}\ (p_C)$ is the partial ELDC including customers' demand (both deterministic and random components), maintenance requirements, and forced outages requirements of the first $N-1$ units in the merit order. Given $\bar{t}_{N-1}\ (p_C)$, the transformation formula enables us to find the corresponding ELDC for an additional unit. By the transformation formula, adding units with zero outage probability on top of the $(N-1)$th unit will not change the function $\bar{t}_{N-1}\ (p_C)$. We can thus refer to this function as the ELDC for the whole system and denote it by $\bar{t}(p_C)$. The expected amount of energy delivered by the last peaking unit in the system is given by the top portion of the ELDC, i.e., by the area under the tail of the curve between p_{ins}^{N-1} and p_{ins}^{N}. The area above p_{ins}^{N} is the expected unserved energy. All the other segments of the curve only indicate the probabilities that the combined demand values are equaled or exceeded, and are irrelevant for our optimal mix considerations of peaking units.

Given the ELDC we will apply the optimal mix procedure to the peak periods in order to find the optimal cutoff point between peaking units and nonsupply and peaking units and intermediate-load units.

We have described the optimal-mix procedure in section 6.2. Here we will apply it to the ELDC rather than the LDC. We denote the various units involved by an index representing their position in the merit order. The last peaking unit will be denoted by unit N and the load-shedding "unit" by unit $N+1$. In accordance with the notation of section 6.2 we define

V_{N-1} = the variable cost of producing a unit of energy (1 kW h) by the last intermediate-load unit,

V_N = the variable cost of producing a unit of energy (1 kW h) by the peaking unit,

V_{N+1} = the social costs of nonsupply per unit of unserved energy,

F_{N-1} = the installation costs (yearly charge) of 1 kW of the last intermediate-load unit,

F_N = the installation costs (yearly charge) of 1 kW of peaking unit.

For all practical purposes we can assume that these costs are inversely related, i.e.,

$$V_{N-1} \le V_N \le V_{N+1},$$
$$F_{N-1} \ge F_N \ge F_{N+1} = 0. \tag{11.1}$$

Also we let p_j and $K_j, j = N - 1, N, N + 1$, denote the installed capacity and the cumulative installed capacity of unit 1 through (and including) unit j.

Assuming that the variable cost per unit of energy produced or curtailed does not vary with the magnitude of the output, the curve describing the total annual investment and operating costs, as a function of the number of hours the unit is scheduled for operation, is given by a straight line with the intercept representing the annual capital cost and the slope indicating the operating costs per hour of operation ($/h). Since the variable and capital costs are inversely related, the curve describing the minimum total expenses as a function of time is a piecewise-concave function as shown in the upper portion of figure 6.2. Thus unit $N + 1$ (nonsupply) will not be economical for more than $\bar{T}(K_N^*)$ hours, unit N (peaking) will be cheaper for up to $\bar{T}(K_{N-1}^*)$ hours, and unit $N - 1$ (last intermediate) will be economical for more than $\bar{T}(K_{N-1}^*)$ hours. The break-even point occurs at the point where the total costs are equal for any two successive units in the merit order, i.e.,

$$F_j + V_j \bar{T}(K_j^*) = F_{j+1} + V_{j+1} \bar{T}(K_{j+1}^*), \tag{11.2}$$

which implies

$$\bar{T}(K_j^*) = \frac{F_j - F_{j+1}}{V_{j+1} - V_j}, \qquad j = N - 1, N. \tag{11.3}$$

The optimal mix is then obtained by projecting the optimal running hours on the ELDC, yielding the values K_j^*, from which we obtain

$$p_j = K_j^* - K_{j-1}^*, \qquad j = N - 1, N. \tag{11.4}$$

The main difficulty in applying the optimal mix algorithm in our case is to assign the appropriate running costs for the nonsupply unit, which from the overall economic point of view should reflect the social cost of unserved energy. We have already mentioned these difficulties. In the absence of reliable predictions of these costs, approximations should be made to solve the optimal mix problem. Clearly, the more accurate are these predictions, the more reliable will be the optimal mix results.

12

The Economic Value of Forced Outages Rates (FOR)

12.1 The Impact of the FOR on the System

Forced outages rates (FOR) for generating units have increased drastically over the years (13). This is due in part to the increase in unit size and to the increased complexity of modern power utilities. Future trends are for yet more increase in the FOR of generating units as power systems grow in size and complexity.

Besides the additional capacity costs incurred by reserve requirements, forced outages of units contribute to an increase in a system's operating costs because of the need to operate machines requiring costlier energy to replace efficient base-load machines that have been forced out of service. In one study it was estimated that unscheduled unit outages may increase production costs by 2 to 5% (28), which, in a system as large as many of the electric utilities, amounts to considerable sums. It is important, therefore, to account for the FOR of units and their impact on the design and performance of power systems in an accurate manner. Schofield and Marsh (30) have indicated three factors which have combined recently to cause electric utilities to take a deeper look at the meaning and cost of forced outages, namely, (a) increases in generating unit size; (b) the importance of differences in reliability between different types of plants; and (c) the financial effect of individual plant reliability.

The incorporation of forced outages into system planning models is not an easy task. For one thing, forced outages are only one factor affecting the reliability and cost performance of a power system; the others include unit size, maintenance requirements, existence of interconnection, reserve margin, shape of load, fluctuation in the demand for power, etc. What makes the problem even more complex is the fact that all these factors are interrelated, with strong interaction existing between the FOR and the other factors. As a result, though it is possible to consider forced outages separately, they should be considered in the context of the whole system, accounting for the other factors simultaneously. Such an overall approach, based on the ELDC, for calculating the net impact of the FOR on the system in reliability and cost terms is presented in this chapter.

We start by analyzing the impact of an increase in the FOR of any given unit on the operating cost and the reliability of the system. Since the FOR can be traded off with the unit size to yield the same level of reliability, a trade-off analysis is carried out in order to balance these two quantities. We then devise procedures for calculating the economic cost of forced outages for three alternatives, meeting the reliability target of the system, namely, (a) adding extra reserve capacity, (b) increasing the unit size, and (c) purchasing power from neighboring companies.

Basically, we will assume that the unit whose FOR is being investigated is a new unit. Assuming a given size and FOR, we first calculate the reliability measures resulting when adding this unit to the system; we then repeat the analysis for the case for which the unit's FOR is increased by, say, 1%, while holding the unit's size constant. The difference in the reliability measures between the two cases will yield the impact of an increase in the FOR of the unit on the reliability performance of the system. If, on the other hand, the FOR of an existing unit is being investigated, we treat it as a "new" unit by removing the unit from the system, and initiating the calculation by generating the resulting ELDC using the procedure described in chapter 8. The same approach is used in determining the impact of the FOR on the cost performance of the system.

We define the *economic* (*marginal*) *cost* of the FOR as the change in the total operating and investment costs of the system resulting from changing the FOR of a given unit by a small increment while maintaining the same reliability level. It is shown that the economic cost of the FOR is strongly dependent on the system's reliability target since it is the attainment of the desired target that actually determines the amount of reserve margin to be maintained in the system.

The marginal cost of the FOR also determines the amount of money and effort worth expending in improving the FOR of a proposed, or a given, unit. While this problem has been dealt with in the past, mainly in a qualitative manner (8), the present chapter offers a quantitative tool for assessing the profitability to the system of changing the FOR of a unit.

The problem of forced outages is also strongly related to maintenance requirements. Rather than change the unit size or the reserve margin level, it is possible to compensate for reduction in reliability resulting from increasing the FOR by rescheduling the maintenance operations. This opens up various possibilities for trade-offs between FOR, unit size, and maintenance schedule, which can also be carried out via the ELDC. However, the impact of changing the maintenance schedule and its relation to the FOR and unit size is beyond the scope of this book.

12.2 Trade-Off between FOR and Loss-of-Load Probability (LOLP)

In the following we assume that the reliability performance of the power system is measured by the *loss-of-load probability* (LOLP) and we calculate the sensitivity of the LOLP to a change in the FOR of a given unit, using the ELDC. When augmenting the system with a unit of capacity p_p MW and outage probability q, the new LOLP of the system, during time periods when the new unit is not under maintenance, is given by

$$\mathrm{LOLP}(q) = (1 - q)\,\bar{t}(p_{\mathrm{ins}} + p_p) + q\,\bar{t}(p_{\mathrm{ins}}), \tag{12.1}$$

where

$\mathrm{LOLP}(q)$ = the LOLP as a function of the outage probability q of the new unit,

$\bar{t}(p)$ = the ELDC evaluated at point p, prior to the addition of the new unit,

p_{ins} = the installed capacity of the system prior to the addition of the new unit.

Suppose now that we increase the FOR of the unit by Δq. The new outage probability will be $q + \Delta q$. Using the above relation, the resulting LOLP, denoted by LOLP $(q + \Delta q)$, is given by

$$\mathrm{LOLP}(q + \Delta q) = (1 - q - \Delta q)\,\bar{t}(p_{\mathrm{ins}} + p_p) + (q + \Delta q)\,\bar{t}(p_{\mathrm{ins}}). \tag{12.2}$$

Taking the difference between the two quantities, we obtain

$$\mathrm{LOLP}(q + \Delta q) = \mathrm{LOLP}(q) + \Delta q[\bar{t}(p_{\mathrm{ins}}) - \bar{t}(p_{\mathrm{ins}} + p_p)]. \tag{12.3}$$

By definition of the ELDC, the quantity in the square brackets is positive, implying LOLP $(q + \Delta q) >$ LOLP (q), an expected result since increased FOR results in lower reliability. It is interesting to note that, assuming all other things are equal, the change in the LOLP resulting from a change in the outage probability of any unit is linear in Δq; an increase Δq in the unit's outage probability increases the LOLP by $\Delta q[\bar{t}(p_{\mathrm{ins}}) - \bar{t}(p_{\mathrm{ins}} + p_p)]$.

The quality of the results depends upon the impact of the maintenance requirements of the unit whose FOR is being investigated on the ELDC. The smaller is the impact, the higher is the accuracy of the resulting LOLP. The results are very accurate, therefore, for peaking units, which do not contribute significantly to "maintenance demand" requirements since it is assumed that they can be maintained in off-peak seasons; they may be somewhat biased if a base unit is added to the system. Yet, it may be argued that if we account for maintenance requirements of all the other units in the system in an accurate manner, then ignoring maintenance requirements of one unit only will not introduce an appreciable error into the calculations. Also, since we are mainly interested in the change in reliability resulting from increasing the FOR, the maintenance requirements of the concerned unit are cancelled anyway in the subtraction process that yielded equation (12.3). But, if more precise results are required, it will be necessary to account for the maintenance requirements of the concerned unit in an accurate manner by examining separately the time periods during which the new unit is under maintenance (assuming then that $q = 1$ and $q + \Delta q = 1$). Clearly, during such periods the LOLP is unchanged, provided that the maintenance schedule of all other units is unchanged.

As an example, consider a power system with $p_{\text{ins}} = 120\%$ of summer peak, and an ELDC as exhibited in figure 3.10. Assuming we augment the system with a unit whose capacity is 10% of summer peak, then any increase in the outage probability of the unit by 1% (i.e., $\Delta q = 0.01$), will increase the LOLP by

$$0.01\,[\bar{t}(120\%) - \bar{t}(130\%)] = 0.01\,(160 - 30) = 1.3 \quad \text{hours}$$

if the new unit is not scheduled for maintenance during the year.

The graph exhibiting the change in the system reliability, as measured by the LOLP, is therefore linear with a slope of 1.3 and an intercept

$$\bar{t}(p_{\text{ins}} + p_p) = \bar{t}(130\%) = 30 \quad \text{hours},$$

as shown in figure 12.1.

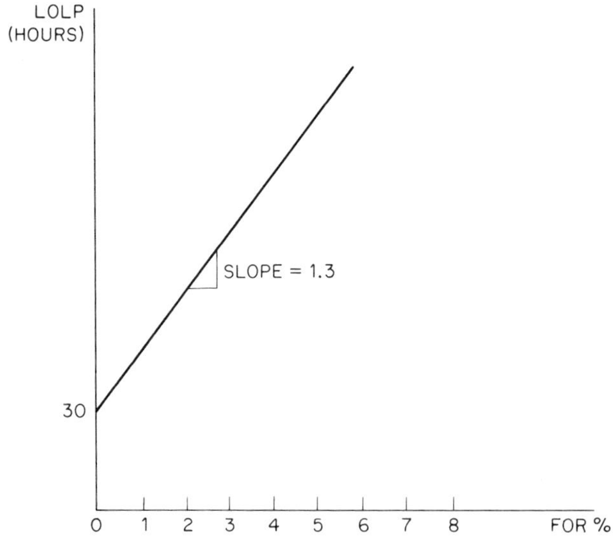

Figure 12.1 The LOLP as a function of the FOR for a unit whose capacity is 10% of summer peak.

12.3 The Impact of the FOR on Operating Costs

The procedure for determining the change in the system's operating costs resulting from a change in the unit's FOR is carried out, as with the LOLP, by augmenting the system by a unit with a given size and varying FOR, and studying the difference between the resulting costs. The task of estimating the operating costs of the system is more complicated than in the case of the LOLP, as it depends on the position of the unit in the system's merit order, the characteristics of the other units in the system, the demand pattern for power, and the availability of units at any given time. While there are many approaches to estimating the operating costs of a power system, the more sophisticated procedures are based on the ELDC. In order to predict the production expenses of a system augmented by one unit, there is a need first to generate the ELDC for the resulting system.

It is shown in chapter 8 that the new ELDC for the augmented system can be estimated in terms of the previous ELDC and the characteristics of the new unit by using a simplified transformation formula,

$$\hat{t}(p_C) = (1 - q)\bar{t}(p_C) + q\bar{t}(p_C - p_p), \tag{12.4}$$

where

$\hat{t}(p_C)$ = the new ELDC evaluated at p_C,

$\bar{t}(p_C)$ = the value of the ELDC at p_C prior to adding the new unit to the system,

p_C = the combined demand for power,

q = the outage probability of the additional unit,

p_p = the generating capacity of the new unit.

Maintenance requirements of the additional unit can be approximated using a slightly modified transformation formula, or else can be incorporated in a precise manner by distinguishing between time periods during which the new unit is under maintenance and time periods during which the new unit is not scheduled for maintenance.

Having derived the new ELDC, either in an exact or an approximate form, the system's operating costs can be estimated either by assuming a merit order loading and using the probabilistic simulation approach, or else by considering the incremental loading procedure and using the more refined calculation method described in chapter 10. The calculation procedure is then repeated for a unit with the same capacity with increased FOR, yielding the resulting change in the operating costs. Unfortunately, it is impossible to express the change in the operating costs in a closed-form formula as with the LOLP. Rather, the difference in the operating costs should be studied numerically by repeating the calculations for various values of FOR. We note that the higher the position of the unit in the merit order, the lower the change in the operating costs resulting from a 1% increase in its FOR. This conclusion can be verfied by remembering that units with a high position in the merit order deliver energy for a small fraction of the time.

12.4 Trade-Off Analysis

It was established earlier that an increase in the FOR of a given unit contributes to an increase in the LOLP of the system, and hence to a decrease in the system's reliability level. A possible alternative, apart from adding extra capacity to the system, to make up for this reduction of reliability is by appropriately increasing the size of the unit involved. In other words, FOR and unit size can be traded off to yield the same level of reliability. The problem of finding the exact increase in unit size necessary to compensate for the reduction in reliability level caused by increasing the FOR by a given percentage is therefore referred to as trade-off analysis. Using the ELDC, the problem amounts to solving for the size of a unit with outage probability $q + \Delta q$ that yields the same reliability level, measured by the LOLP, as for a unit with capacity p_p MW and a lower outage probability q.

Adding a unit with capacity p_p MW and outage probability q, we have

$$\text{LOLP} = (1 - q)\bar{t}(p_{\text{ins}} + p_p) + q\bar{t}(p_{\text{ins}}). \tag{12.5}$$

Adding a unit with capacity p_p' MW and outage probability $q + \Delta q$, we obtain

$$\text{LOLP} = (1 - q - \Delta q)\bar{t}(p_{\text{ins}} + p_p') + (q + \Delta q)\bar{t}(p_{\text{ins}}). \tag{12.6}$$

The unknown capacity p_p' is obtained by equating the two equations:

$$\bar{t}(p_{\text{ins}} + p_p') = \frac{(1 - q)\bar{t}(p_{\text{ins}} + p_p) - q\bar{t}\,(p_{\text{ins}})}{1 - q - \Delta q}. \tag{12.7}$$

From this p_p' is derived by referring to the corresponding ELDC. By varying the calculations for various values of Δq, the trade-off curve between the outage probability (or, equivalently, the FOR) and the unit size for any reliability level is derived.

As an example we refer to the system whose ELDC is exhibited in figure 3.1 assuming a LOLP reliability target of 20 hours a year. We investigate what combinations of unit size and outage probability yield the desired reliability level by applying the above formula to a unit whose outage probability varies in the range 2–10%. As before, we assume that the system's installed capacity is 120% of summer peak. The results are exhibited in table 12.1 and illustrated in figure 12.2.

Table 12.1 Calculation of the Trade-Off Curve

q (%)	$\bar{t}(120 + p_p')$ (hours)	p_p' (% of summer peak)
2	133.5	13.5
3	133.9	13.9
4	134.5	14.5
5	135.2	15.2
6	136.0	16.0
7	137.0	17.0
8	138.2	18.2
9	139.5	19.5
10	141.0	21.0

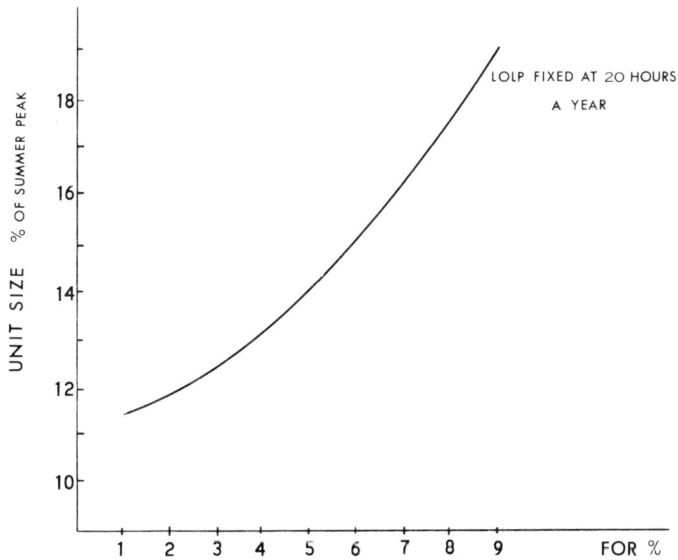

Figure 12.2 Trade-off curve between FOR and unit size for LOLP = 20 hours (example).

It can be seen that the increment in unit size increases with the increase of the outage probability since larger and larger units are required to compensate for any additional increase in the outage probability. It is very important to note the differences in unit size for various levels of outage probabilities; for example, an increase of the outage probability from 9 to 10% is equivalent, from a reliability point of view, to increasing the unit size from 19.5 to 21% of summer peak, i.e., an increase of 7.7%. It can also be seen that the percentage of increment in unit size increases as the outage probability gets larger. The same conclusions can also be drawn from the trade-off curve itself. The instantaneous rate of change in the unit size for any increase in the outage probability is given by the slope of the trade-off curve at any point.

As previously discussed, the accuracy of the trade-off results depends upon the impact on the ELDC of the maintenance requirements of the unit whose characteristics we are investigating. It should be noted that the maintenance requirements depend upon the unit's size. This is not accounted for in our example since we assume no maintenance of the new unit and no change in the maintenance schedule of the existing units. In an actual case the maintenance schedule, including that of the new unit, must be given in order to obtain an accurate trade-off curve.

Given the trade-off curve, the operating costs associated with each point on the curve can also be determined. For any combination of unit size and outage probability yielding the same level of reliability we first have to derive the resulting ELDC either in an approximate or an exact form, and then apply the operating cost algorithm to calculate the associated production expenses. The results can be exhibited in a graph describing the change in the system's operating costs for various combinations of outage probabilities and unit sizes yielding the same level of reliability. The procedure can be extended to a cost–reliability analysis by repeating the calculations for various reliability levels.

12.5 The Economic Cost of the FOR

The *economic cost*, or, alternatively, the *marginal cost*, of the FOR will be defined as the change in the total costs resulting from changing the FOR of a unit by 1% while maintaining the same level of reliability. Two components are usually included in the marginal cost of the FOR (in fact, in the marginal cost of any parameter of the system), namely, capacity cost and production cost. Since the amount of installed capacity to be maintained in the system depends upon the desired reliability level, the marginal cost of the FOR is strongly dependent on the system's design target. In calculating the economic cost of the FOR it is necessary, therefore, to analyze the impact of increasing the FOR on the reliability performance of the system, and to compensate for the resulting reduction in reliability either by adding extra reserve capacity to the system or else by increasing the unit size; both involve additional capacity costs and result in different operating costs. The economic cost of the FOR, therefore, depends upon the particular approach selected to meet the reliability target of the system.

The problem of finding the amount of reserve needed for meeting a given reliability target has been dealt with in chapter 9. For all practical purposes we can assume that peaking units are used to provide for the extra reserve capacity required to compensate for the decreased level of reliability resulting from increasing the unit's FOR. This assumption is particularly valid in our case because we are dealing with marginal, and therefore very small, amounts of reserve capacity. With this assumption, the procedure for determining the extra reserve level for meeting the required reliability target turns out to be simple. In particular, it is assumed that adding a peaking unit to the system, provided it is working only a small fraction of the time, does not change the ELDC of the system; this assumption gives a good approximation since peaking units do not contribute significantly to requirements for maintenance and forced outages. The level of installed capacity required to meet a given level of LOLP is obtained, therefore, by reading its value directly off the ELDC. If p_{ins} is the required installed capacity, we actually solve

$$\text{LOLP} = \bar{t}(p_{\text{ins}}), \tag{12.8}$$

where $\bar{t}(p_{\text{ins}})$ is the ELDC evaluated at point p_{ins}.

With this method, the capacity cost component of the economic cost of the FOR will be equal to the extra annualized cost of peaking units required to compensate for the reduced reliability level caused by increasing the unit's FOR by 1%. More accurately, assuming we augment the system with a unit with capacity p_p MW and outage probability q, we first derive the ELDC for the augmented system in either an exact or approximate form, and use the above approach to determine the additional investment in peaking units required to maintain the reliability of the system at the desired level. We then repeat the analysis for a unit with the same capacity and an FOR increased by 1%. The difference between the two additional annual capacity costs will yield the capacity cost component of the economic cost of the FOR. The generation cost component is obtained in a similar manner using the procedure outlined in section 13.3. Given the ELDC for the augmented system, we calculate the operating cost of the system by using either an approximate or an exact algorithm. We then repeat the calculation for a unit with a 1% increase in its FOR. The difference between the two operating cost figures yields the additional generation costs that can be attributed to the change in the FOR. The sum of the capacity and generation components of the cost constitutes the extra annual cost incurred by increasing the FOR of the unit by 1%. By definition, this is equal to the economic cost of the FOR.

In case the reserve requirements are to be met by increasing the unit's size rather than by added peaking capacity, we have to resort to the trade-off curve to evaluate the capacity and generation component of the economic cost of the FOR. Specifically, we ask ourselves what change in unit size is required to compensate for the decreased reliability level resulting from increasing the FOR by 1%. This figure is readily obtained from a trade-off curve like figure 12.2. The annual cost of this extra capacity constitutes the capacity cost component of the economic cost of the FOR. The operating cost component is obtained in a manner similar to the one described above by studying the corresponding difference in the production costs. The economic cost of the FOR is then obtained as the sum of the capacity cost and the generation cost components.

A third possibility for meeting the desired reliability level arises in systems operating in a pool environment. In such a case, rather than adding capacity costs to compensate for the decrease in reliability resulting from increasing the FOR, the power company may meet the desired reliability level by purchasing the deficient energy from members of the power pool. Considering this extra source of energy as an additional "unit" placed on top of the last unit in the system's merit order, the expected amount of energy to be bought during a year in order to meet the reliability target is equal to the area under the ELDC between the system's installed capacity and the desired LOLP level. Then the economic cost of the FOR will be equal to the cost the power company pays for the extra energy required to compensate for the decrease in reliability resulting from increasing the FOR of the unit by 1%.

By carrying out the calculations for various reliability levels, we can derive a curve representing the economic cost of FOR as a function of system reliability for any of the three alternatives described above.

Schofield and Marsh (30) discuss the impact of changing the FOR of any given unit on the company's earnings per share. Such an analysis can obviously be carried out in a more effective way with the more refined procedures for calculating the economic cost of the FOR that have been presented in this chapter.

12.6 Implications for Decision Making

Several implications of the procedures described in this chapter for decision making can be pointed out. First and foremost, the above analysis is extremely useful in determining the exact effect of the FOR and the unit size of any unit under consideration for addition to the system. Such an analysis is particularly important for bid evaluation, when it is necessary to assess the impact of a variety of units offered by various manufacturers in order to focus on the fine details of the type and make of the units to be installed in the system. And conversely, for units currently existing in the system, the capability of accurately generating the economic costs of the FOR is an important feature in deciding when to retire a unit and replace it with a new one. Besides its effect on the reliability of the system, forced outages also bear a substantial impact on the cost of the system, as well as on the prices customers have to pay for electricity. An increase in the FOR of a unit contributes to an increase both in the capacity and in the production costs of the system, which, if sustained, may increase electricity costs for the consumer. Price considerations, and their relation to the economic cost of FOR, may also be a factor in capacity expansion or retirement decisions.

13

Rate Structure Analysis by ELDC—Marginal Capacity Costs

13.1 Introduction

In this and the next chapter a procedure for calculating time-dependent expected marginal costs is suggested. This procedure constitutes an effective tool for supporting peak-load pricing decisions. Although there might be many difficulties in translating marginal costs into actual prices, marginal considerations should be the starting point for any time-differentiated pricing scheme. As stated in part I, beyond the variation of costs due to the changing pattern of demand it is necessary to account also for the random elements of both the supply and the demand, which have a substantial impact on the cost of meeting an increment of energy at any hour. The major contribution of these chapters is in devising a computational method for calculating time-dependent marginal costs under conditions of supply-and-demand uncertainty, both for energy and capacity. Failure to make appropriate allowances for these uncertainties may result in a simplistic pricing scheme that will not guarantee recovery of the costs incurred in power generation.

Several important elements, related to rate structure analysis, can be analyzed by the ELDC:

- operating cost predictions and estimation of marginal energy charges;
- the impact on reliability, and on cost of shaving peak demand; and
- determining peak responsibility and allocating capacity costs.

Operating cost calculations are important components in determining the energy charge for covering the variable operating cost of electric power generation. For a "minimum- cost" system this energy charge is equal to the marginal cost of providing electricity by the most inefficient unit in the system (the last unit in the merit order) (24). A reliable estimate of the marginal energy costs requires an accurate description of the behavior of the system during the peak period, and a refined procedure for calculating the operating cost of the last unit in the merit order—both are available through the ELDC.

The impact of shaving peak demand on reliability and on cost is relevant to decisions regarding incentives given peak customers in order to get them off during peak hours. The determination of this impact falls under the category of marginal price analysis, which is best carried out with the help of the ELDC.

A very important factor in rate structure analysis, especially with regard to peak-load pricing, is the problem of peak responsibility. According to conventional approaches, only customers that consume electricity during peak hours should be held responsible for peak demand, and should be charged with the extra cost incurred in satisfying peak demand. However, since the installed capacity is a function of the LOLP, and the LOLP is affected by all types of customers, not just peak customers, it is evident that off-peak costomers also bear responsibility, though smaller, for peak demand. It will be shown how the ELDC is used for determining peak responsibility in systems exposed to forced outages.

13.2 Marginal Capacity Costs and the Design Target

The problem of apportioning marginal capacity costs is intimately related to the design target employed in planning for capacity expansion of the system since it is the attainment of this target that determines the level of the installed capacity to be added to the system. To find the share of capacity cost to be borne by each hour, it is necessary to determine the impact of the load at any given hour on the design target of the system, and then to assign the marginal capacity cost accordingly.

Taking the LOLP as an example of a reliability design target, we first establish the contribution of each hour to the LOLP in a formal manner by investigating the contribution of the load at any given hour to the LOLP. We then devise an algorithm to determine this contribution accurately for each hour of the year. It is shown that the contribution of each hour to the LOLP is equal to its curtailment probability, and is therefore larger for peak hours than for off-peak hours. In particular, if the installed capacity is less than the peak demand for power, each hour of consumption during peak periods contributes a full hour to the LOLP. This contribution is somewhat lower in shoulder periods, and approaches zero, but is still positive, at base-load hours. A sustained increase in demand at any hour will therefore call for adding extra capacity to the system; otherwise the LOLP design target will not be met. Consequently, each hour of the year, including off-peak hours, makes some contribution to the incurrence of capacity costs, and should be charged with capacity cost in relation to its associated curtailment probability. By casting the analysis in terms of utilization factors, we are able to interpret the curtailment probability associated with each hour as the expected fraction of time the last kilowatt of installed capacity is scheduled for operation at that hour, thus providing an economic justification for charging customers with capacity costs relative to the curtailment probability at the hour of consumption. Provisions for incorporating random deviations of customers' demand and maintenance requirements are included.

While being only one component in the cost of electricity, marginal capacity costs are nevertheless a major component, especially in view of the fact that electric power utilities are highly capital-intensive systems. Variable load pricing schedules which fail to make allowance for loss-of-load probability are simplistic and give rise to pricing differentials between peak and off-peak periods of excessive amounts.

13.3 The Effect of the Load in a Given Hour on the LOLP

This section formally establishes the contribution of each hour to the loss-of-load probability, ignoring, in the first stage, maintenance requirements of the generating units and uncertainties in customers' demand. The basic conclusion is as follows:

In a power system in which the generating units are subject to forced outages, each hour of the year, including off-peak hours, contributes theoretically a positive amount, however small, to the LOLP of the system. Furthermore, this contribution is equal to the curtailment probability associated with the hour.

We prove this statement by determining the impact of removing the load in a given hour, i.e., reducing it to zero, on the LOLP. Figure 13.1 describes the load–duration curve (LDC) of an electric power system. Recall that forced outages are thought of as an additional type of "demand," in fact the self-demand of the power company for installed capacity. Three possible outage levels are superimposed on the LDC in figure 13.1, with each outage raising the ELDC in a parallel manner at a distance equal to the forced-out capacity level. The occurrence probabilities of the three outage levels are also shown in the figure, and are to be explained further later. Assuming the standard two-state model for each generating unit due to forced outages (operative or nonoperative), and assuming that the outage probabilities of the various units are independent of one another, the probabilities of the capacity in outage are drawn from a multinomial distribution in which the outage probabilities of the individual units are parameters. At any given time, the sum of customers' demand for power and the capacity in outage, referred to as the *combined demand*, exhibits the total requirements for generating capacity for satisfying customers' demand and the "demand" for forced outages. Denoting the combined demand by p_C, and the expected number of hours in a given period during which the combined demand for power equals or exceeds p_C by $\bar{t}(p_C)$, we have

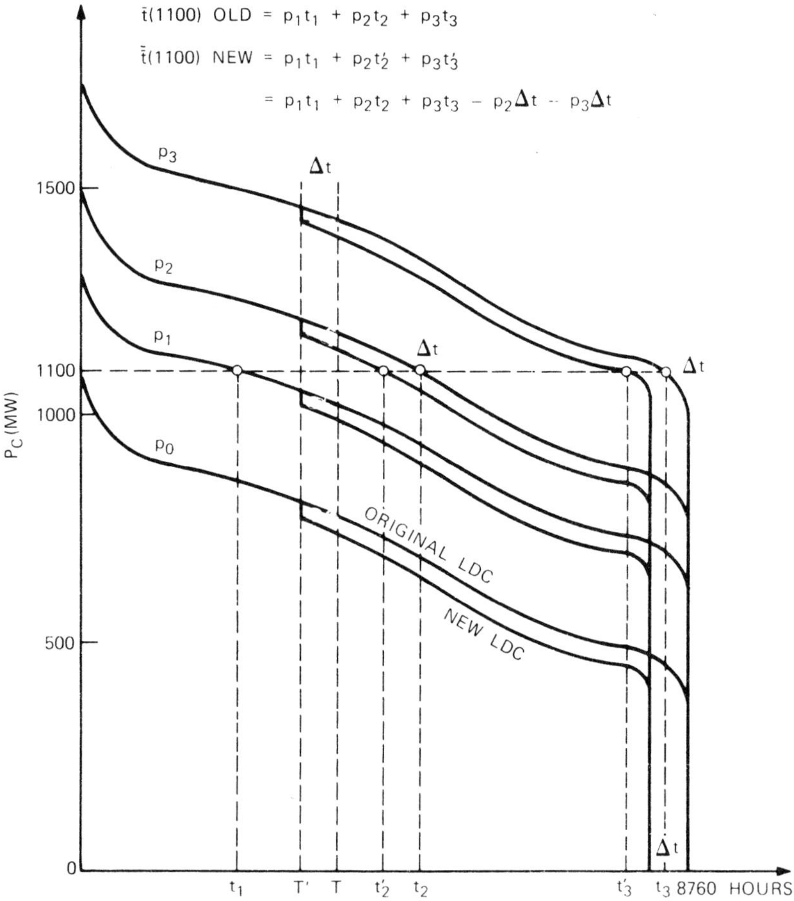

Figure 13.1 Finding the effect of the load of given hour on the equivalent LDC and the LOLP.

$$\bar{t}(p_C) = \sum_{S} P(p_F)\, t(p_C - p_F), \tag{13.1}$$

where

$P(p_F)$ = the probability of failures at a level p_F,

$t(x)$ = the LDC evaluated at point x,

S = the set of all events of up and down units for which $p_F \leq p_C$.

The above formula is actually an alternative form for expressing the convolution of unit outages with customers' load. By definition, equation

(13.1) is a partial ELDC containing only two components of demand, namely, customers' demand and forced outages requirements.

Recall that

$$\text{LOLP} = \bar{t}(p_{\text{ins}}) = \sum_{S} P(p_{\text{F}})\, t(p_{\text{ins}} - p_{\text{F}}), \tag{13.2}$$

where p_{ins} is the installed capacity of the system.

Assuming that we reduce customers' load at time T to zero for a duration of time Δt, the LDC in fiigure 13.1 will be shifted Δt to the left in the interval $(T, 8760)$ while it will remain the same in the interval $(0, T - \Delta t)$. The same happens to the three curves combining customers' load and forced outages, as is seen in the figure. Denoting the new ELDC by $\bar{\bar{t}}(p_{\text{C}})$, we have

$$\bar{\bar{t}}(p_{\text{C}}) = \sum_{S_1} P(p_{\text{F}})\, t(p_{\text{C}} - p_{\text{F}}) + \sum_{S_2} P(p_{\text{F}})[t(p_{\text{C}} - p_{\text{F}}) - \Delta t], \tag{13.3}$$

where

$\bar{\bar{t}}(p_{\text{C}})$ = the value of the new ELDC at point p_{C},

S_1 = the set of all events of up and down units for which $p_{\text{F}} \leq p_{\text{C}} - p_{\text{D}}(T)$, with $p_{\text{D}}(T)$ denoting customers' demand on the LDC at time T,

S_2 = the set of all events of up and down units for which $p_{\text{F}} > p_{\text{C}} - p_{\text{D}}(T)$.

In fact, the summation in equation (13.1) is actually split into two sets of elements, those which are not affected by the change in customers' demand in the interval Δt and those which are affected by the change in demand. Figure 13.1 demonstrates how $\bar{\bar{t}}(1100)$ is calculated.

By definition, the new LOLP is obtained by

$$\bar{\bar{t}}(p_{\text{ins}}) = \sum_{S_1} P(p_{\text{F}})\, t(p_{\text{ins}} - p_{\text{F}}) + \sum_{S_2} P(p_{\text{F}})[t(p_{\text{ins}} - p_{\text{F}}) - \Delta t]. \tag{13.4}$$

The difference between the LOLPs before and after the change in customers' demand at time T is therefore

$$\Delta\,\text{LOLP} = \bar{t}(p_{\text{ins}}) - \bar{\bar{t}}(p_{\text{ins}}) = \sum_{S_2} P(p_{\text{F}})\Delta t; \tag{13.5}$$

this result is directly obtained recalling that Δt is the probability of customers' load equaling $p_{\text{D}}(T)$. $\sum_{S_2} P(p_{\text{F}})$ is by definition the probability of an outage level exceeding $p_{\text{ins}} - p_{\text{D}}(T)$, thus causing a shortage. The product of these probabilities is the contribution of Δt to the LOLP.

When Δt is very small, we obtain in the limit

$$\frac{d\text{LOLP}}{dt} = \sum_{S_2} P(p_F). \tag{13.6}$$

Since $P(p_F) > 0$ regardless of the value of T, $d\text{LOLP}/dt > 0$ for all values of T, thus establishing the first part of our statement that each hour of the year contributes a positive amount to the LOLP of the system.

The set S_2 is actually made up of all combinations of up and down units that lead to a shortage situation and the summation on the right-hand side of equation (13.6) is the curtailment probability associated with time T, as already mentioned. It therefore follows from equation (13.6) that the contribution of each hour to the LOLP and the curtailment probability associated with that hour are identical, as was claimed by the second part of our opening statement.

It is easy to see that the curtailment probability gets larger as we move toward peak hours. Logically, the probability of loss of load should be greater during peak hours than during off-peak hours. Mathematically, as we approach peak hours the set S_2 contains more and more combinations of up and down units, resulting in more terms contributing to the summation in equation (13.6) and therefore to larger curtailment probabilities. In the extreme case where T is the peak hour, and the installed capacity is less than the peak demand, S_2 will contain all combinations of up and down units, so that $d\text{LOLP}/dt = 1$. This is an expected result since in such a case an elimination of one hour from the peak will reduce the LOLP by exactly one hour. The contribution to LOLP is somewhat lower in the succeeding periods, sometimes referred to as the *shoulder periods* (23), and approaches zero, but is still positive, in base-load periods. Even though the contribution of each hour to the LOLP declines as we get further from the peak, it is substantially high during shoulder and even during off-peak periods, indicating that a substantial portion of the hours of the year, not just peak hours, has an impact on the LOLP attained by the system.

Having established our basic statement that all hours of the year contribute a positive amount to the LOLP, it follows that each hour of the year, including off-peak hours, should bear some responsibility for the incurrence of capacity costs. In the following section we will suggest an approach to apportioning capacity costs based on the curtailment probabilities associated with the hours of the year. But we will first devise a better method for calculating these probabilities.

13.4 Measuring the Contribution of Each Hour to the LOLP

Equation (13.6) provides us with a very inefficient means of calculating the curtailment probability associated with each hour since it requires that the occurrence probabilities of all combinations of up and down units that lead to a shortage situation be calculated—which, in a system with N units, may contain as many as 2^N combinations. In this section we shall devise a more efficient approach to calculating curtailment probabilities based on an alternative expression for calculating the LOLP. We shall demonstrate the algorithm with an example.

In figure 13.2 we represent our three-unit system example with capacities 250 MW, 300 MW, and 400 MW, and outage probabilities 0.06, 0.04, and 0.02, respectively, with forced outages superimposed on the LDC. Eight combinations of up and down units are possible in this case, each of which is shown in the figure along with its occurrence probability. Table 3.1 demonstrated the calculation process of these probabilities (see chapter 3).

To find the LOLP in the figure, we draw a horizontal line through the installed capacity p_{ins}, which cuts the various curves at times $t(p_{\text{ins}} - p_{\text{F}j})$, $j = 0, 1, 2, \ldots, 2^N - 1$. By the definition of the LOLP the following relation holds:

$$\text{LOLP} = \sum_{j} t(p_{\text{ins}} - p_{\text{F}_j})\, P(p_{\text{F}_j}), \tag{13.7}$$

where $p_{\text{F}_j}, j = 0, 1, \ldots, 2^N - 1$, are the various possible outage levels in increasing order of magnitude, each of which corresponds to a given combination of up and down units, with p_{F_0} denoting the situation of no outages, i.e., $p_{\text{F}_0} = 0$, and $P(p_{\text{F}_j}), j = 0, 1, \ldots, 2^N - 1$, are the corresponding occurrence probabilities.

By expanding the right-hand side of equation (13.7), expressing each $t(p_{\text{ins}} - p_{\text{F}_j}), j = 0, 1, \ldots, 2^N - 1$, as a summation of consecutive time intervals, we obtain

$$\text{LOLP} = t(p_{\text{ins}}) + \sum_{j=1}^{2^N - 1} [t(p_{\text{ins}} - p_{\text{F}_j}) - t(p_{\text{ins}} - p_{\text{F}_{j-1}})]\left[1 - \sum_{i=1}^{j} P(p_{\text{F}_{i-1}})\right]. \tag{13.8}$$

Equation (13.8) offers an alternative expression for calculating the LOLP. Basically it expresses the LOLP as a summation of areas of rectangles, each with a base given by the difference in value of $t(p_{\text{ins}} - p_{\text{F}_j})$, at two successive values of p_{F_j}, and a height expressed in terms of outage probabilities. The first term in equation (13.8) can also be interpreted as an area of a rectangle with base $t(p_{\text{ins}})$ and height 1. The expression is illustrated graphically at the lower part of figure 13.2. In this figure the $2^3 = 8$ outage levels are shown, where $p_{\text{F}_0} = 0$, $p_{\text{F}_1} = 250, \ldots, p_{\text{F}_8} = 950$, and $P(p_{\text{F}_0}) = 0.8844, \ldots, P(p_{\text{F}_7}) = 0.00005$.

Denoting the height of the rectangle at any given hour i by h_i, equation (13.8) can be expressed in hours as

$$\text{LOLP} = \sum_{i=1}^{8760} h_i; \tag{13.9}$$

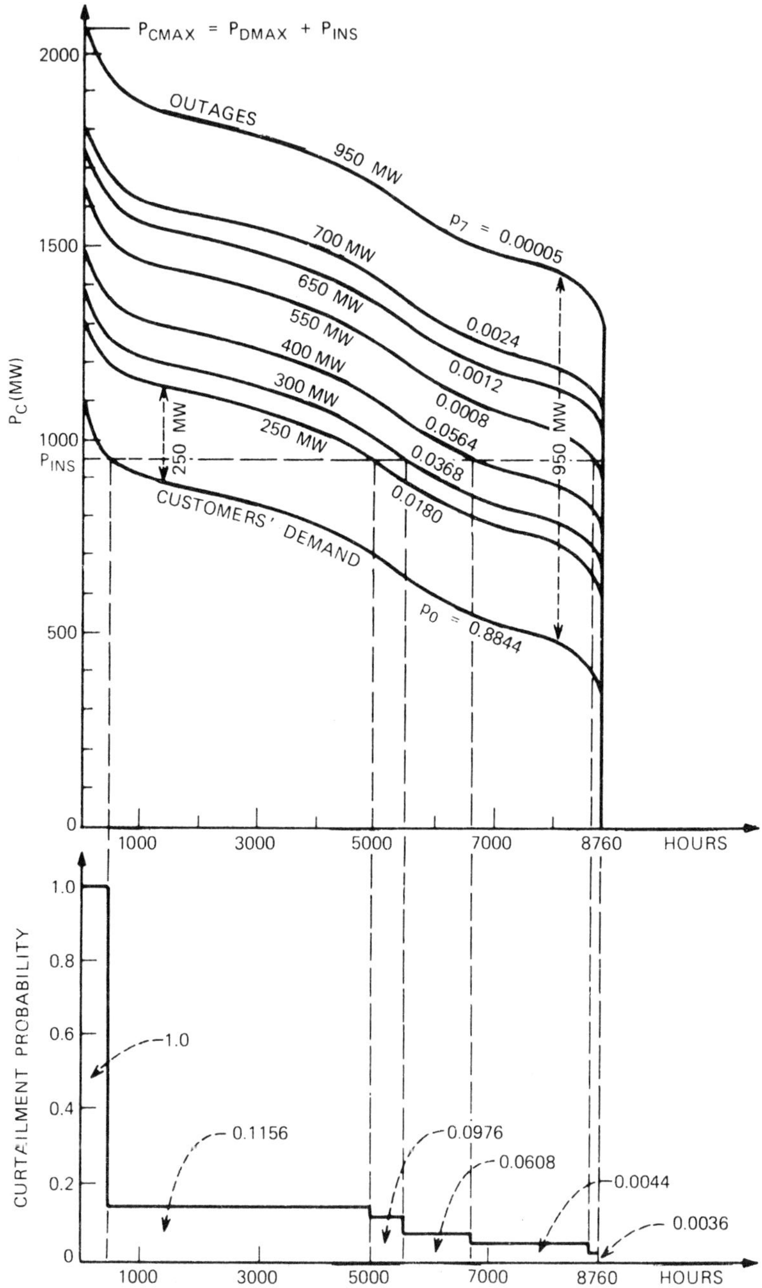

Figure 13.2 Determining the contribution of each hour to the LOLP for a system in which the installed capacity is less than the peak demand.

h_i is obviously the contribution of hour i to the LOLP. It is easily seen from equation (13.8) that h_i is equal to the curtailment probability associated with hour i since it is obtained as a summation of occurrence probabilities of all events (combinations of up and down units) which lead to a shortage situation. Note that

$$h_i = 1 - \sum_{k=0}^{j} P(p_{F_{k-1}}) = \sum_{i=j+1}^{2^N - 1} P(p_{F_{k-1}}).$$

The curve at the lower part of figure 13.2 therefore defines the curtailment probability associated with each hour of the year, and will consequently be termed the curtailment probability curve. The sum of curtailment probabilities over all hours of the year is actually equal to the LOLP of the system.

Since the outage probabilities in a power system are discrete, the curtailment probability curve is actually a discontinuous step function, as shown in figure 13.2, with 2^N points of discontinuities, N being the number of units in the system. While it is very easy to derive the curtailment probability curves for systems containing very few units, it seems almost impossible to generate the curve for larger systems since the number of the discontinuous points rises exponentially with the number of units in the system. As N get larger, the curtailment probability curve approaches a continuous pattern since both the width and the height of the steps become smaller and smaller. It is therefore possible to approximate the curtailment probability curve, even for relatively small systems, by evaluating the curtailment probabilities at only a few discontinuous points (failure levels) and fitting a continuous curve through these points. In this way the task of estimating the curtailment probability curve, especially for large, diversified systems, is greatly simplified. Typical curtailment probability curves are exhibited in figure 13.3.

In addition to providing a convenient computational means, the curtailment probability curve offers a significant insight into the behavior of the curtailment probabilities over the hours of the year as a function of the installed capacity. It will be useful, in this connection, to distinguish between two cases. In the first case described in figure 13.2, which is rather hypothetical and given for demonstration only, the installed capacity is smaller than the peak demand for power. As suggested by the figure, the curtailment probability associated with the hours in the interval $0 \leq t(p) \leq t(p_{ins})$ is 1 since any demand in this interval will surely result in some level of shortage. The curtailment probability associated with the subsequent period where $t(p_{ins}) < t(p) \leq t(p_{ins} - p_{F_1})$ is 0.1156, which is indeed smaller, but definitely not negligible. In fact, the total contribution of the hours in the second interval is larger than the contribution of the hours at the interval of peak hours. By equation (13.8) the contribution of

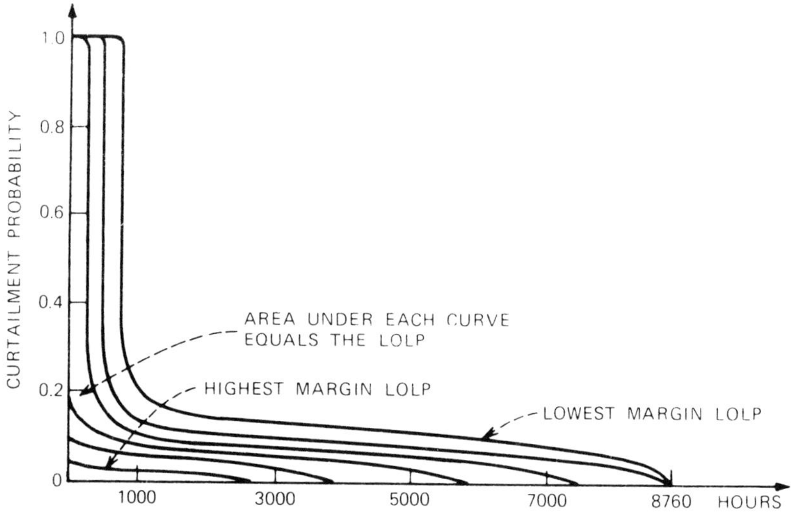

Figure 13.3 Curtailment probability curves.

an interval to the LOLP is obtained by multiplying the length of the interval in hours by the associated curtailment probability. Then the contribution of the first interval to the LOLP is $500 \times 1.0 = 500$ hours, and of the second interval $4500 \times 0.1156 = 520$ hours. Even in this case, where peak demand exceeds the installed capacity, peak customers are not the only contributors to the LOLP.

In the second case we assume the real situation in which the installed capacity exceeds the peak demand for power. We modify the previous example to reflect this situation by lowering all the curves in figure 13.2 by 200 MW in a parallel manner. The situation is illustrated in figure 13.4. We note that the curtailment probability associated with the peak hours has dropped to 0.1156—a sharp decline from the previous case—which results in a more even distribution of the curtailment probability curve. In fact, it can be shown that in identical generating systems, the larger is the excess of the installed capacity over peak demand, the more even will be the curtailment probability curve. Since most electric power utilities possess excess capacity over peak demand, it is plainly wrong to charge only peak customers with capacity costs; rather, the distribution of these costs should be more even to reflect the fact that the curtailment probabilities are more uniformly distributed along the hours.

Of course, this does not imply that all customers should be charged the same. The off-peak customer will always have smaller responsibility than the on-peak customer—smaller, but theoretically not zero. Actual computation can show, however, that, say, after 3000 hours the curtailment

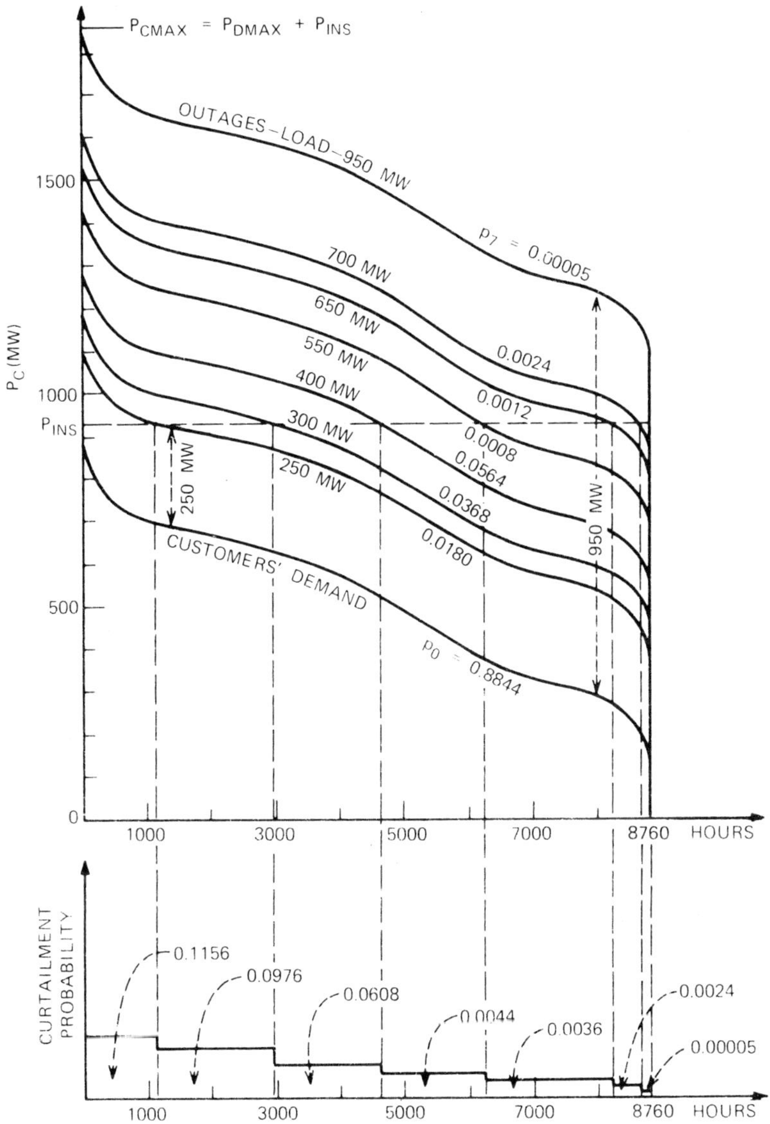

Figure 13.4 Determining the contribution of each hour to the LOLP for a system possessing excess capacity over peak demand (example).

probabilities decline by three orders of magnitude. This is due to the very small probability of the event that a sufficiently large number of units will be out simultaneously, which is what it takes to create shortage at that level of lower demand. The main conclusion is therefore that shoulder customers should be charged. Having established this basic conclusion, we now proceed to devise an appropriate procedure for apportioning capacity costs among various customers.

13.5 Assigning Marginal Capacity Costs

Optimal mix considerations suggest that in power systems with many different kinds of capacity with different construction costs, the marginal capacity cost for the system equals the construction cost of the last kilowatt of peaking capacity, i.e., the cost of the last unit in the merit order per kilowatt (17). Denoting the annual fixed charge of this capital cost by F, the capacity cost of the last kilowatt-hour of operation per year is given by dividing the fixed charge of capital by the number of hours the last kilowatt is scheduled for operation during the year. We note that the expected number of hours at which the last machine in the merit order runs at maximum capacity equals the LOLP of the system (expressed in hours), when assuming zero outage probability for this machine. But this is also equal to the expected time duration the last kilowatt will be scheduled for operation in the given year. The capacity cost of the last kilowatt of installed capacity per hour of operation per year is thus given by F/LOLP. This is the marginal capacity cost for the system which we have to apportion between the various types of customers.

Given that each hour of the year contributes differently to the need to expand the system, a logical approach to apportioning marginal capacity costs is to charge each hour relative to its contribution to the incurrence of these capacity costs as given by its contribution to the LOLP, or, equivalently, by its curtailment probability, i.e.,

$$C_i = \frac{F}{\text{LOLP}} h_i, \tag{13.10}$$

where

C_i = the marginal capacity costs chargeable to hour i,

h_i = the curtailment probability associated with hour i.

In a realistic price scheme these capacity costs should clearly be reflected in the price that each customer pays for electricity.

To offer yet a stronger economic justification for apportioning the capacity costs relative to the curtailment probabilities, we cast the analysis in terms of utilization factors of the last kilowatt of installed capacity.

Because of the random fluctuation of the installed capacity due to forced outages, there is always a possibility, however small, that the last unit will be called into operation at full capacity at any hour of the year, including off-peak hours. However, the probability with which the last kilowatt will be called upon to generate power at hour i equals the curtailment probability associated with the hour.

It is also noteworthy that the curtailment probability associated with hour i is equal to the expected fraction of time the last kilowatt is scheduled for operation at that hour. The proof is derived by looking at the problem in terms of the LOLP. If the capacity of the last unit is reduced by a kilowatt, the LOLP will be increased by the fraction of time the last kilowatt is scheduled for operation at that hour. But by definition, the latter is exactly the contribution of the hour to the LOLP, which is, in turn, equal to the curtailment probability associated with the hour. Hence, by apportioning capacity costs relative to curtailment probabilities, we actually charge each hour according to the utilization it makes of the last kilowatt of installed capacity. Recalling that the marginal capacity costs of the system are measured in terms of construction costs of peaking capacity, this approach to allocating capacity costs has economic merit. Indeed, the curtailment probability associated with hour i can be thought of as the utilization responsibility of the hour. Consequently, it provides a sounder economic basis for charging customers with capacity costs than the traditional peak responsibility methods.

As an example of this calculational procedure, we consider the situation illustrated in figure 13.4 and assume an annual capacity charge of \$28/kW/year. The calculations are exhibited in table 13.1. (The LOLP obtained in this example is clearly not a realistic one, and therefore not a representative figure. Electric utilities are generally designed to meet far

Table 13.1 Assigning Marginal Capacity Costs

(1) Interval (hours)	(2) Number of hours in interval	(3) Curtailment probability (h_i)	(4) Contribution of each interval to the LOLP (2) × (3)	(5) $\frac{F}{\text{LOLP}} h_i$ (¢/kW/year)
0–1100	1100	0.1156	127.2	0.75
1100–3000	1900	0.0976	185.4	0.63
3000–4700	1700	0.0608	103.4	0.40
4700–6300	1600	0.0044	7.0	0.03
6300–8100	1800	0.0036	6.5	0.02
8100–8700	600	0.0024	1.5	0.02
8700–8760	60	0.00005	0	0
	8760		LOLP = 431.0	

lower levels of LOLP. This example is chosen only to demonstrate the principles of the calculations.)

It is possible to extend this approach to account also for the different consumption patterns of the various sectors (residential, commercial, industrial, agricultural, etc.). Denoting by l_{ij} the percentage of load due to sector j at hour i, the marginal capacity cost attributable to each hour is given by $C_i l_{ij}$. The latter should then be included as part of the charge for the energy consumed by sector j at hour i.

Even though we were able to express the theoretical principles of capacity cost allocation rather explicitly, there are important practical complications that might limit the application of the above approach in the real world, at least in the first stages of implementing the variable load pricing scheme. Beyond various qualifications which relate to peak-load pricing in general, there are other problems more peculiar to capacity cost allocation. First we should mention the problem of measuring the marginal capacity cost in a system with many different kinds of capacity with different construction costs. We have indeed assumed, as is usually conceded, that in a perfect system, in which capacity is continuously adjusted to meet the demand at minimum cost, the marginal cost of an additional kilowatt of capacity is the construction cost of a peaking unit. In reality, however, various factors make a truly minimum-cost system unattainable at any point in time, so that basing marginal capacity cost on construction costs of peaking units may lead to biased results. Another complication is concerned with the size of the incremental unit that is relevant for measuring marginal capacity cost, especially since the addition of capacity in a power system takes place in large blocks. Finally, because the lifetime of a power plant is relatively long and because economic conditions are dynamically changing, the problem of determining the annual charges of capital is also a complicated one.

The question of identifying rating periods is another important part of variable load pricing schemes. There is no sense in assigning a different price to each hour of the year; rather, the hours should be grouped into rating periods, with a different price for each. However, uncertainties about demand and the fluctuation of the available installed capacity at any given time make the problem of defining the rating periods a difficult one. Nevertheless, once the rating periods have been established, the weighted average contribution of each sector to the incurrence of capacity costs should be calculated by summing its contributions for all hours of the period, and then charging capacity costs by sectors rather than by hours. The disadvantage of this method is that it does not allocate capacity costs according to the instantaneous time of consumption as suggested by peak-load pricing schemes. However, the approach is the only practical one and can be implemented quite easily.

Another point of interest is concerned with optimal mix considerations. Most studies on price schemes are related to minimum cost systems, for which capacity is perfectly adjusted to meet the demand for electric power. However, it is evident from the discussions and the analysis in this book that the problem of assigning marginal capacity costs to various hours, as distinguished from the process of measuring these capacity costs, has little to do with optimal mix considerations. Once these costs are established, the only factor that actually determines the share of each hour in recovering the marginal capacity costs is the design target of the system. In a perfectly adjusted system, where the reserve capacity is set at a level where marginal capacity costs are equal to the expected marginal curtailment costs, the LOLP will also exhibit the optimal level of nonsupply for the system. However, in most power systems, the cost to society of curtailment is usually unknown, and the level of LOLP is determined rather arbitrarily—for example, according to a criterion such as "1 day in 10 years." Such systems are clearly not optimal. In fact, in a recent study (32), it is shown that a planning criterion with a "1-day-in-10-years" loss-of-load probability as its reliability target may be unecomonically and unjustifiably high. The fact that the marginal capacity costs facing the utility may not be optimal from the national point of view does not, however, affect the procedure described above of apportioning the capacity costs to the various customers.

13.6 The Effects of Maintenance and Random Deviations of Demand

Maintenance requirements have a substantial effect on the level of reserve required to meet a desired level of LOLP. In particular, maintenance requirements increase the contribution of off-peak hours to the LOLP because most maintenance operations, especially of large units, are usually scheduled at off-peak hours. As a result, we may sometimes expect higher probabilities for shortages during off-peak hours compared to the corresponding probabilities for shoulder and even peak periods. It is essential, therefore, to include maintenance requirements in the process of assigning marginal capacity costs. Because of the dependence of the forced outages levels on the maintenance schedule, it is first necessary to partition the period involved into subperiods of constant maintenance and to arrange the load in each subperiod into a load–duration curve, with maintenance requirements considered as part of the demand for power. The process of assigning marginal capacity costs to the various hours is then carried out with respect to each subperiod separately by using the procedure previously described. The results for the various hours are then combined to yield each hour's share in recovering the marginal capacity costs. The difficulties with this approach are two. First, this procedure results in an increased volume of calculation and a need to estimate accurately maintenance requirements in the long run and their schedule. Second, it is arguable whether off-peak customers should assume the full responsibility for the increase in the curtailment probability resulting from scheduling maintenance requirements during off-peak hours. Clearly, were it not for these hours, maintenance operations would have been scheduled for periods of higher demand, making it necessary to add extra reserve capacity to meet the system reliability target. Therefore, off-peak hours actually contribute to saving capacity costs which would otherwise have been incurred if maintenance operations were to be scheduled during periods other than off-peak hours. The question, then, is whether off-peak hours should be given some credit for making it possible to meet maintenance requirements without adding extra reserve capacity, and if so, how large should this credit be and what should be the additional charges to be imposed on peak consumption in return. We have not resolved this problem and merely observe that it requires further investigation.

Also, the preceding analysis was only concerned with supply uncertainties as reflected by forced outages requirements. A further refinement of the procedure might be attempted by taking cognizance of demand uncertainties. This can be done by incorporating random deviations of customers' demand into the LDC by means of the convolution process, as explained in chapter 8.

Rate Structure Analysis by ELDC—Marginal Energy Costs

14.1 Graphical Presentation

By the manner of its construction, the LDC is a monotonically decreasing function, implying a one-to-one correspondence between the customers' demand p_D and the time $t(p_D)$. Hence, an alternative approach to calculating time-differentiated expected marginal energy costs is to calculate marginal energy costs related to each demand level first, and then attribute each cost to the corresponding hour, based on the LDC. To do this we need to determine what units assume the role of the last supplying unit for each demand level and with what probabilities. An approach based on the combined demand for power will be suggested for carrying out this calculation.

The concept of the combined demand provides us with a convenient means for determining for any unit the range of demands for which it may be used as the last supplying unit, along with the corresponding occurrence probabilities. For this purpose, we will define the partial combined demand p_C^i as the sum of customers' demand and forced outages requirements of the first i units in the merit order, i.e.,

$$p_C^i = p_D + \sum_{k=0}^{2^i-1} p_{F_k} P(p_{F_k}), \tag{14.1}$$

where p_D is customers' demand and p_{F_k} and $P(p_{F_k})$ are the 2^i ordered outage levels of the first i units and their associated occurrence probabilities, respectively.

Being a function of random variables, each partial combined demand p_C^i is a random variable too, assuming as many as 2^i different values, with a probability function denoted by $P_i(p_C)$. In deriving these probability functions for the various values of i, we will use the recursive relation given by equation (8.17) to relate the probability functions of two successive variables p_C^{i-1} and p_C^i:

$$\begin{aligned} P_i(p_C) &= (1-q_i)P_{i-1}(p_C) + q_i P_{i-1}(p_C - p_i), \qquad i = 1, 2, \ldots, N, \\ P_0(p_C) &= f(p_C), \end{aligned} \tag{14.2}$$

where

$f(p_D)$ = the probability density function of customers' load, evaluated at p_D,

p_i = the capacity (in MW) of unit i,

q_i = the outage probability of unit i,

N = the number of units in the system.

We note that the original recursive equations of Booth (9) are similar to equations (14.2) with p_F substituting for p_C in the argument, and with $P_0(p_F) = 1$ for $p_F = 0$ and 0 otherwise. These equations have been used previously in studies of reliability analysis as a means of deriving the capacity in outage distribution of a system augmented by one unit in terms of the capacity of outage distribution of the system prior to adding the new unit and the characteristics of the additional unit. We have also used this formula to derive successively the complementary density functions of the partial combined demand, referred to as the *partial ELDC*, and denoted by $\bar{t}_i(p_C)$ in chapter 8.

We denote by p^i_{ins} the total installed capacity of the first i units in the merit order i.e.,

$$p^i_{\text{ins}} = \sum_{k=1}^{i} p_k. \qquad (14.3)$$

Then, by definition, any excess of the partial combined demand over p^{i-1}_{ins} will be supplied by unit i as the last unit, provided it is working, up to the capacity p^i_{ins}. In other words, unit i is the last supplying unit whenever

$$p^{i-1}_{\text{ins}} \le p^i_C \le p^i_{\text{ins}}. \qquad (14.4)$$

When $p^{i-1}_{\text{ins}} \ge p^i_C$, unit i is not scheduled for generation at all since the first $i - 1$ units in the merit order are capable of satisfying all the demand for power; and conversly, when $p^i_C \ge p^i_{\text{ins}}$, unit i is indeed scheduled for operation, but not as the last unit (except for the case where the unit is also the last unit in the merit order).

Each value of p^i_C can be arrived at from various values of p_D, with various probabilities, depending upon the combinations of up and down units. To find the contribution of each unit i to the expected marginal energy cost of a given demand level, it is therefore necessary to find the probability of "transition" from p_D to any of the values of p^i_C satisfying inequality (14.4). In order to derive these probabilities, we will use the recursive equation (14.2) in a successive manner.

Assume first that there are no generating units in the system. Then, by definition, the partial combined demand p^0_C will be equal to p_D, customers' load, with probability 1. The function relating p^0_C to p_D will therefore have the shape of a straight line emanating from the origin at a 45° angle, as shown in figure 14.1A.

Adding one unit to the system, the partial combined demand p_C will assume the following probability distribution:

$$p_C^1 = \begin{cases} p_D & \text{with probability } 1 - q_1 \\ p_D + p_1 & \text{with probability } q_1. \end{cases} \tag{14.5}$$

The graph exhibiting the relation p_C^1 to p_D will now be composed of two straight lines, as shown in figure 14.1B, namely, the original line of figure 14.1A with probability $(1 - q_1)$ and a parallel line at a distance of p_1 with probability q_1.

Adding a second unit to the system, the distribution of p_C^2 in terms of p_D will assume four possible values:

$$p_C^2 = \begin{cases} p_D & \text{with probability } (1 - q_1)(1 - q_2) \\ p_D + p_1 & \text{with probability } q_1(1 - q_2) \\ p_D + p_2 & \text{with probability } (1 - q_1)q_2 \\ p_D + p_1 + p_2 & \text{with probability } q_1 q_2; \end{cases} \tag{14.6}$$

the graph relating p_C^2 to p_D will contain four straight lines, parallel to each other, each taking place with a different occurrence probability as demonstrated in figure 8.1C.

In general, given the lines relating p_C^{i-1} to p_D and their probabilities, and applying the recursive relation, the family of lines relating p_C^i to p_D will be composed of the original lines, with the corresponding probabilities each multiplied by $(1 - q_i)$, and a new set of lines, obtained by raising the original lines in a parallel manner at a distance p_i, with the associated probabilities multiplied by q_i. The curves are exhibited in figure 14.2. We note that each straight line in figure 14.2 corresponds to a given combination of up and down units so that the number of possible lines in the figure is 2^i.

Recalling the condition expressed by the inequality in (14.4), it is clear that each line in figure 14.2 which appears in the range $p_{\text{ins}}^{i-1} \le p_C^i \le p_{\text{ins}}^i$ refers to a certain combination of up and down units in which unit i is loaded to generation as the last unit. A given value of p_D (such as p_{D_1}) may be intersected by more than one straight line in the above interval, indicating that several combinations of up and down units result in unit i being the last supplying unit for that demand level. And conversely, if a value of p_D is not intersected by any line in the above interval (such as p_{D_3}), it means that unit i can never be the last supplying unit for that demand level.

To find the expected marginal energy cost for a demand level p_D it is necessary to express the partial combined demand p_C^i in terms of p_D, as suggested by figure 14.2, for all i, $i = 1, 2, \ldots, N$, and then to apply

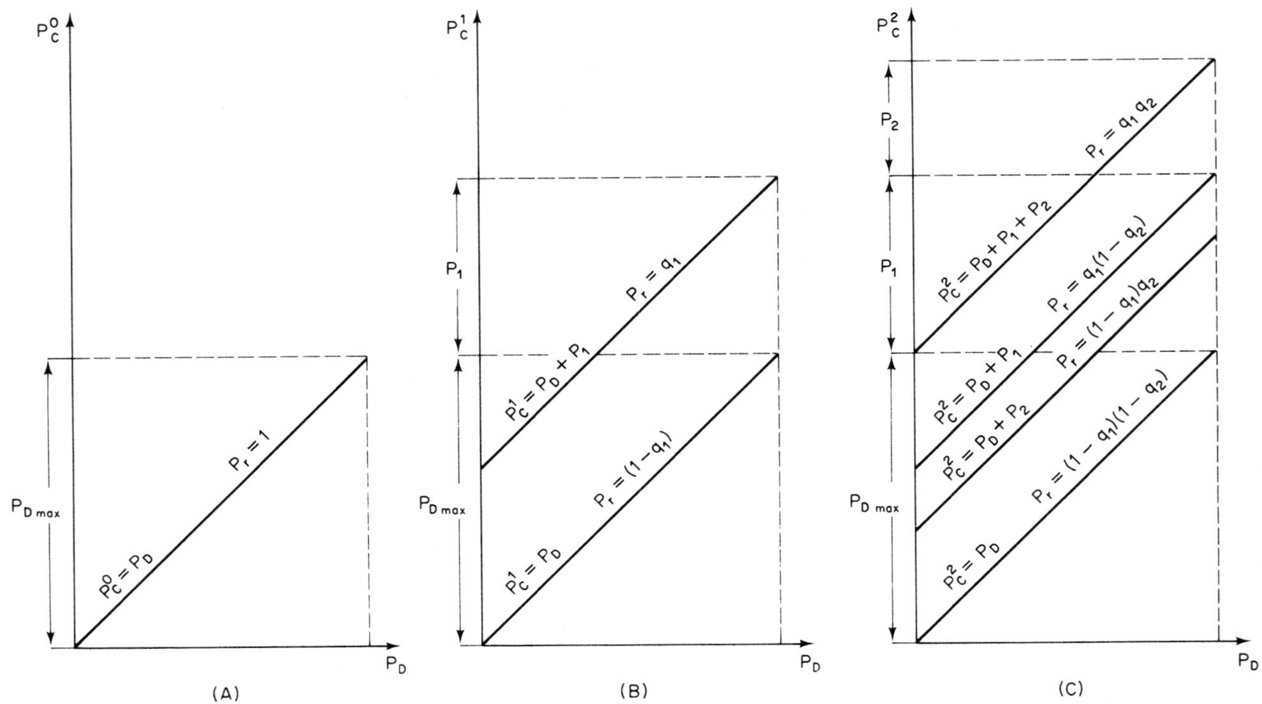

Figure 14.1 P_C^i as a function of P_D for a two-unit system.

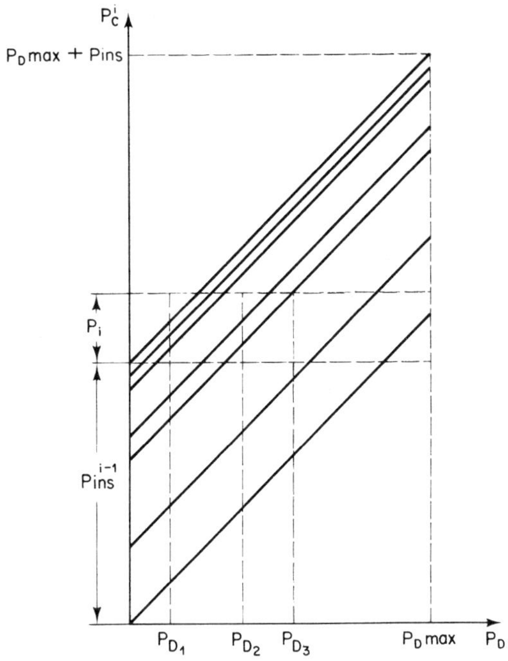

Figure 14.2 P^i_C as a function of P_D in the general case.

condition (14.4) in order to find the combinations in which unit *i* serves as the last supplying unit. Summing the probabilities associated with these combinations (lines in the figure) provides us with the probability that unit *i* serves as the last unit when the demand level is p_D. Multiplying the resulting probability by the fuel cost of unit *i* and summing over all values of *i* yields the expected marginal energy cost at demand level p_D.

As an example of the process, we assume a three-unit system with capacities 300 MW, 200 MW, and 100 MW, outage probabilities q_1, q_2, and q_3, average fuel costs of c_1, c_2 and c_3 (¢/kW), respectively, and a peak demand ($p_{D\max}$) of 500 MW. The 300-MW unit is assumed to be the first unit in the merit order, the 200-MW unit the next, and the 100-MW unit the last. We demonstrate the calculation graphically in figures 14.3, 14.4, and 14.5, adding successively one unit at a time and calculating, at each stage, the contribution of unit i, $i = 1, 2, 3$, to the demand-related expected marginal energy cost. The total expected marginal energy cost for each demand level p_D is obtained by summing up the contributions of all the units. As mentioned, this marginal cost can then be attributed to the corresponding hour $t(p_D)$ on the LDC. Note that since unit *i* must be operative whenever it supplies the marginal energy, it is possible to

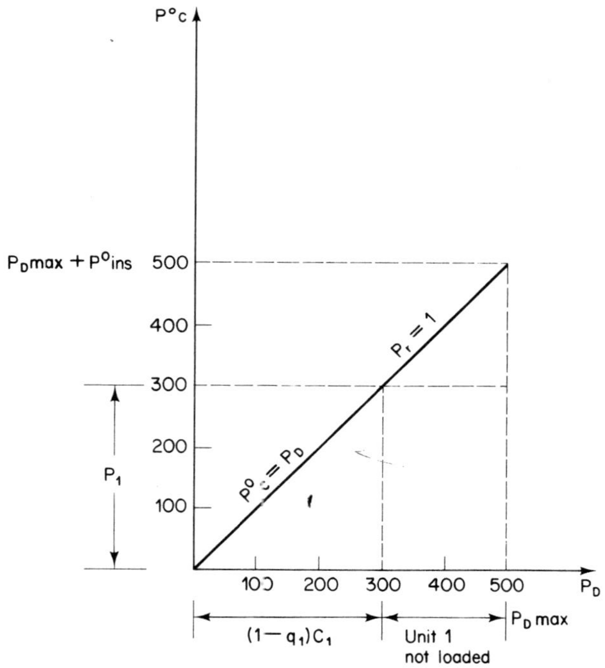

Figure 14.3 Contribution of unit 1 to the expected marginal energy cost for every demand level.

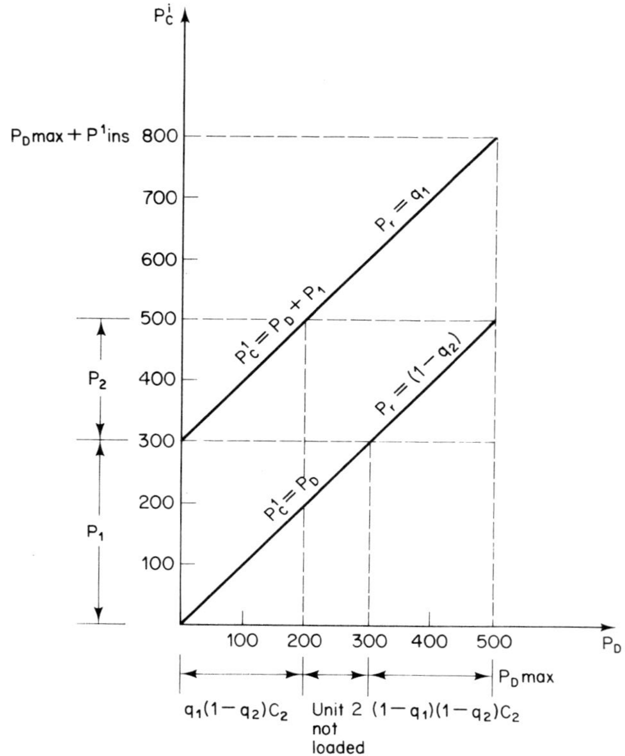

Figure 14.4 Contribution of unit 2 to the expected marginal energy cost for every demand level.

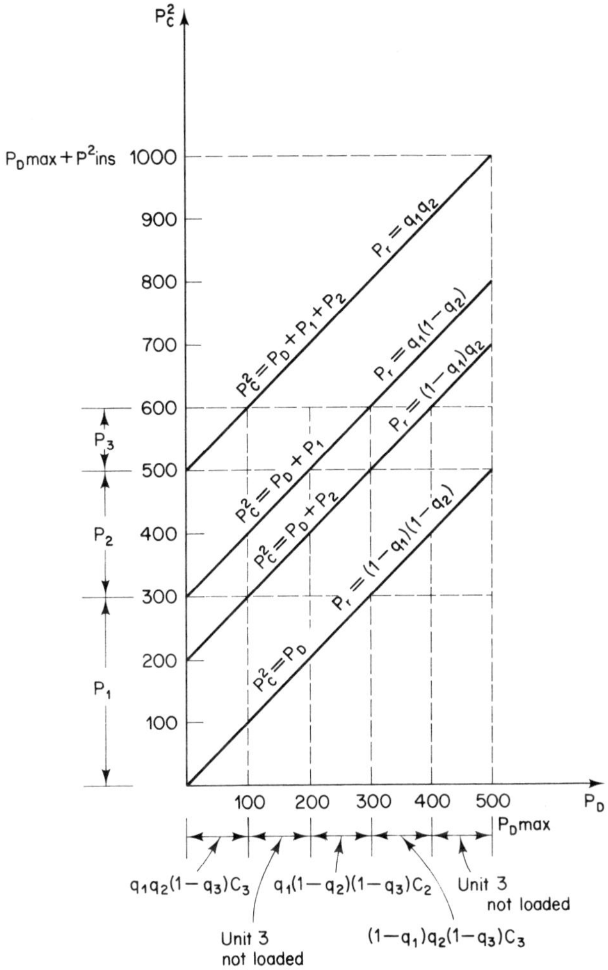

Figure 14.5 Contribution of unit 3 to the expected marginal energy cost for every demand level.

calculate its contribution to the expected marginal energy cost by using the graph describing p_C^{i-1}, rather than p_C^i, as is done in our example. This eliminates the need to draw the graph of p_C^3 in the example.

14.2 Mathematical Formulation

The approach suggested in the preceding sections will now be formulated as a mathematical algorithm.

Let B^i be the transition probability matrix from a partial combined demand p_C^{i-1} to a partial combined demand p_C^i. Note that B^i has the dimension of $p_{D\max} + p_{\text{ins}}^{i-1}$ rows and $p_{D\max} + p_{\text{ins}}^{i}$ columns, where $p_{D\max}$ denotes the peak demand for power. By the recursive relation (14.2) there are two possible transitions from p_C^{i-1} to p_C^i: either $p_C^i = p_C^{i-1}$ with probability $(1 - q_i)$, or $p_C^i = p_C^{i-1} + p_i$ with probability q_1. Hence the matrix B^i is of the form

$$B^i = \begin{bmatrix} 1-q_i & 0 & \dots & 0 & q_i & & & & 0 \\ & & 1-q_i & 0 & \dots & 0 & q_i & & \\ 0 & & & & 1-q_i & 0 & \dots & 0 & q_i \end{bmatrix}, \qquad (14.7)$$

with $(1 - q_i)$ elements on the main diagonal, q_i elements on the off-diagonal at a distance of p_i columns, and zeros elsewhere. The initial matrix B^0, the transition probability matrix from p_D to p_C^0, is the identity matrix with dimensions $(p_{D\max} \times p_{D\max})$.

Similarly, we denote by A^i the transition probability matrix from p_D to p_C^i, with dimensions $p_{D\max}$ rows and $p_{D\max} + p_{\text{ins}}$ columns. By the definition of A^i and B^i, the following relation holds:

$$A^i = A^{i-1} B^i, \qquad i = 1, 2, \dots, N, \qquad (14.8)$$

where $A^0 = B^0$.

Following the discussion in the last section, the columns of A^i in the interval $p_{\text{ins}}^{i-1} \le p_C^i \le p_{\text{ins}}^i$ contain the probabilities of transition from various levels of P_D to those values of p_C^i for which unit i is called upon to generate power as the last unit. Then, the weighted contribution of the ith unit to the demand-related expected marginal energy cost, denoted by V_j^i, is obtained by

$$V_j^i = \sum_k a_{jk}^{i-1}(1 - q_i)c_i, \qquad j = 1, 2, \dots, p_{D\max}, \qquad (14.9)$$

where a_{jk}^{i-1} is the (jk)th element of the matrix A^{i-1}, and where the summation is carried out over all elements of A^{i-1} for which $p_{\text{ins}}^{i-1} \le p_C^i \le p_{\text{ins}}^i$. The total expected marginal energy cost associated with each demand level, denoted by V_j, is therefore obtained as a summation of all the individual contributions:

$$V_j = \sum_{i=1}^{N} V_j^i, \qquad j = 1, 2, \dots, p_{D\max}. \qquad (14.10)$$

It may be more convenient to express the above formulas in matrix notation. Letting

$\bar{A}^{i-1}$ = the submatrix of matrix A^{i-1} of dimensions $(p_{D\max} \times p_i)$ containing those columns of A^{i-1} for which $p_{\text{ins}}^{i-1} \le p_C^i \le p_{\text{ins}}^i$,

Q^i = the $(p_i \times 1)$ vector with all elements equaling $(1 - q_i)c_i$,

V^i = the $(p_{\mathrm{D\,max}} \times 1)$ vector containing the contribution of unit 1 to the expected marginal energy cost for demand level j,

V = the total expected marginal energy cost for demand level j,

we have

$$V^i = \bar{A}^{i-1} Q^i, \qquad i = 1, 2, \ldots, N, \tag{14.11}$$

where V is given as a vector summation:

$$V = V^1 + V^2 + \ldots + V^N. \tag{14.12}$$

As an example of the calculation process we refer again to the example dealt with in figure 14.5. For purposes of calculation we will work in pu (per unit) notation, by defining 100 MW = 1 pu. Thus we have

$$B^0 = \begin{bmatrix} 1 & 0 & 0 & 0 & 0 \\ 0 & 1 & 0 & 0 & 0 \\ 0 & 0 & 1 & 0 & 0 \\ 0 & 0 & 0 & 1 & 0 \\ 0 & 0 & 0 & 0 & 1 \end{bmatrix},$$

$$B^1 = \begin{bmatrix} 1-q_1 & 0 & 0 & q_1 & 0 & 0 & 0 & 0 \\ 0 & 1-q_1 & 0 & 0 & q_1 & 0 & 0 & 0 \\ 0 & 0 & 1-q_1 & 0 & 0 & q_1 & 0 & 0 \\ 0 & 0 & 0 & 1-q_1 & 0 & 0 & q_1 & 0 \\ 0 & 0 & 0 & 0 & 1-q_1 & 0 & 0 & q_1 \end{bmatrix},$$

$$B^2 = \begin{bmatrix} 1-q_2 & 0 & q_2 & 0 & 0 & 0 & 0 & 0 & 0 & 0 \\ 0 & 1-q_2 & 0 & q_2 & 0 & 0 & 0 & 0 & 0 & 0 \\ 0 & 0 & 1-q_2 & 0 & q_2 & 0 & 0 & 0 & 0 & 0 \\ 0 & 0 & 0 & 1-q_2 & 0 & q_2 & 0 & 0 & 0 & 0 \\ 0 & 0 & 0 & 0 & 1-q_2 & 0 & q_2 & 0 & 0 & 0 \\ 0 & 0 & 0 & 0 & 0 & 1-q_2 & 0 & q_2 & 0 & 0 \\ 0 & 0 & 0 & 0 & 0 & 0 & 1-q_2 & 0 & q_2 & 0 \\ 0 & 0 & 0 & 0 & 0 & 0 & 0 & 1-q_2 & 0 & q_2 \end{bmatrix},$$

$$A^0 = B^0, \qquad A^1 = A^0 B^1 = B^1, \qquad A^2 = A^1 B^2,$$

$$A^2 = \begin{bmatrix} (1-q_1)(1-q_2) & 0 & (1-q_1)q_2 & q_1(1-q_2) \\ 0 & (1-q_1)(1-q_2) & 0 & (1-q_1)q_2 \\ 0 & 0 & (1-q_1)(1-q_2) & 0 \\ 0 & 0 & 0 & (1-q_1)(1-q_2) \\ 0 & 0 & 0 & 0 \end{bmatrix}$$

$$\left.\begin{matrix} 0 & q_1q_2 & 0 & 0 & 0 & 0 \\ q_1(1-q_2) & 0 & q_1q_2 & 0 & 0 & 0 \\ (1-q_1)q_2 & q_1(1-q_2) & 0 & q_1q_2 & 0 & 0 \\ 0 & (1-q_1)q_2 & q_1(1-q_2) & 0 & q_1q_2 & 0 \\ (1-q_1)(1-q_2) & 0 & (1-q_1)q_2 & q_1(1-q_2) & 0 & q_1q_2 \end{matrix}\right] .$$

By definition of the submatrices $\bar{A}^i$ we have

$$\bar{A}^0 = \{\text{columns of } A^0 \text{ corresponding to } 0 \le p_C^1 \le 300 \text{ MW}\}$$

$$= \begin{bmatrix} 1 & 0 & 0 \\ 0 & 1 & 0 \\ 0 & 0 & 1 \\ 0 & 0 & 0 \\ 0 & 0 & 0 \end{bmatrix},$$

$$\bar{A}^1 = \{\text{columns of } A^1 \text{ corresponding to } 300 < p_C^2 \le 500 \text{ MW}\}$$

$$= \begin{bmatrix} q_1 & 0 \\ 0 & q_1 \\ 0 & 0 \\ 1-q_1 & 0 \\ 0 & 1-q_1 \end{bmatrix},$$

$$\bar{A}^2 = \{\text{columns of } A^2 \text{ corresponding to } 500 \le p_C^3 \le 600 \text{ MW}\}$$

$$= \begin{bmatrix} q_1q_2 \\ 0 \\ q_1(1-q_2) \\ (1-q_1)q_2 \\ 0 \end{bmatrix}.$$

Therefore

$$V^1 = \bar{A}^0 \begin{bmatrix} (1-q_1)c_1 \\ (1-q_1)c_1 \\ (1-q_1)c_1 \end{bmatrix} = \begin{bmatrix} (1-q_1)c_1 \\ (1-q_1)c_1 \\ (1-q_1)c_1 \\ 0 \\ 0 \end{bmatrix},$$

$$V^2 = \bar{A}^1 \begin{bmatrix} (1-q_2)c_2 \\ (1-q_2)c_2 \end{bmatrix} = \begin{bmatrix} q_1(1-q_2)c_2 \\ q_1(1-q_2)c_2 \\ 0 \\ (1-q_1)(1-q_2)c_2 \\ (1-q_1)(1-q_2)c_2 \end{bmatrix},$$

$$V^3 = \bar{A}^2 \,[(1 - q_3)c_3] = \begin{bmatrix} q_1 q_2 (1 - q_3) c_3 \\ 0 \\ q_1 (1 - q_2)(1 - q_3) c_3 \\ (1 - q_1) q_2 (1 - q_3) c_3 \\ 0 \end{bmatrix},$$

and

$$V = V^1 + V^2 + V^3 = \begin{bmatrix} (1 - q_1) c_1 + q_1 (1 - q_2) c_2 + q_1 q_2 (1 - q_3) c_3 \\ (1 - q_1) c_1 + q_1 (1 - q_2) c_2 \\ (1 - q_1) c_1 + q_1 (1 - q_2)(1 - q_3) c_3 \\ (1 - q_1)(1 - q_2) c_2 + (1 - q_1) q_2 (1 - q_3) c_3 \\ (1 - q_1)(1 - q_2) c_2 \end{bmatrix}.$$

It is easily seen that the results obtained correspond exactly with those obtained graphically in figure 14.5.

An additional useful property of the vectors V^i will now be discussed. For every level of demand some unit must be the last unit supplying the marginal energy or else a shortage situation exists. For each level of demand j the probability that unit i will be loaded as the last unit is given by $(1/c_i)\, V^i_j$. Therefore for every level j of demand the shortage probability is given by

$$1 - \sum_{i=1}^{N} \frac{1}{c_i} V^i_j.$$

By referring to the LDC, these probabilities can be attributed to the corresponding hours $t(p_D)$. In the previous chapter we have referred to these probabilities as the curtailment probabilities associated with the hours of the year, and have shown that they constitute the basis for charging customers with capacity costs in power systems planned to meet a given LOLP reliability design target. Hence, in addition to providing the expected marginal energy costs associated with each hour, the above calculation approach also provides us, as a by-product, with the basic data required to assign marginal capacity costs to various hours of consumption.

In our example,

$$\sum_{i=1}^{3} \frac{1}{c_i} V^i = \begin{bmatrix} (1 - q_1) + q_1 (1 - q_2) + q_1 q_2 (1 - q_3) \\ (1 - q_1) + q_1 (1 - q_2) \\ (1 - q_1) + q_1 (1 - q_2)(1 - q_3) \\ (1 - q_1)(1 - q_2) + (1 - q_1) q_2 (1 - q_3) \\ (1 - q_1)(1 - q_2) \end{bmatrix},$$

and therefore the curtailment probabilities are

$$1 - \frac{1}{c_i} V^i = \begin{bmatrix} q_1 q_2 q_3 \\ q_1 q_2 \\ q_1 q_2 + q_1 q_3 - q_1 q_2 q_3 \\ q_1 + q_2 q_3 - q_1 q_2 q_3 \\ q_1 + q_2 - q_1 q_2 \end{bmatrix}.$$

It is easy to sec that the proposed computational procedure results in substantial savings in the volume of calculations required for deriving the expected marginal energy costs for each hour, as compared with the enumeration method.

To recall, as many as 2^N elements are required to calculate expected marginal energy cost for each hour in the enumeration method. Assuming we repeat the calculations for M hours, we end up with $M \times 2^N$ calculations, a number which increases exponentially with the number of units in the system. The bulk of the computational effort in the proposed approach is in deriving the matrices $A^i, i = 1, 2, \ldots, N$. This involves N multiplications of matrices whose dimensions increase as we proceed to add units to the calculation process. Suppose that we calculate the expected marginal energy costs for M values of p_D (which will then be attributed to M hours on the LDC) for a system containing 20% reserve. Then by definition $p^N_{\text{ins}} = 1.2\, p_{\text{Dmax}}$, so that the matrix A^i of maximum dimensions, i.e., A^N, will contain M rows and 2.2 M columns (recall that the dimentions of A^N are $[p_{\text{Dmax}} \times (p_{\text{Dmax}} + p_{\text{ins}})]$). Hence, instead of dealing with 2^N events for M hours in the enumeration method, we are dealing with N products of matrices of maximum dimensions ($M \times 2.2M$) in the proposed method. The savings in computations are evident.

14.3 Incorporating Maintenance Requirements and Random Deviations of Customers' Load

To attempt further accuracy in calculating the expected marginal energy costs for each hour of consumption, it is necessary to account for maintenance requirements and random deviations of customers' load. In doing so, we follow the approach used in deriving the ELDC. This approach was also used in assigning marginal capacity costs to various hours. The period involved is divided into subperiods of constant maintenance. The expected marginal energy cost is calculated separately for each subperiod and then aggregated for the whole period.

To incorporate the random element of customers' demand we obtain first the complementary density function of the sum of the deterministic component of customers' load and the random component of customers' load. The resulting curve is actually a partial ELDC with no provisions for forced outages requirements. In accordance with our previous notation, we denote it by $\bar{t}_0(p_C)$ (8).

Given $\bar{t}_0(p_C)$ for each subperiod of constant maintenance, we then calculate the expected marginal energy cost for each subperiod separately, using the same procedure as described in the preceding section, by substituting $\bar{t}_0(p_C)$ for the regular LDC, avoiding the units under maintenance. We note that since the LDC enters the calculation process only in the last stage, when translating demand-related marginal cost into the corresponding time, the calculation process used is not affected by the introduction of random deviations of demand. The number of computations, however, increases linearly with the number of subperiods of constant maintenance.

Appendix
The Simulation Program

The simulation program is made up of a main routine and several subroutines (see block diagram), as described in the following.

A The Main Routine

The main routine reads in the input data and prints it out. It also prints out some of the more important output reports. The various input items that are read by the main routine are as follows:

- the number of years for which the simulation is to be run;
- the number of times to repeat the subroutine that generates random breakdowns each week;
- the number of generating units in the basic system;
- the vector of prices of conventional fuel for each geographic region;
- the prices of fuel for gas turbines and for nuclear units;
- the base date, i.e., the date starting from which the simulation is to be performed;
- the vector of weekly average demands for the planning period;
- description of typical load–duration curves;
- a table that identifies which of the typical load–duration curves to associate with each week of the year;
- the description of the existing system of generating units;
- the number of investment alternatives that are to be investigated;
- the descriptions of the generating units in the investigated investment alternatives;
- the maintenance schedule of the generating system for the planning period [1]

[1] A later version is equipped with an additional subroutine for generating its own maintenance schedule on the basis of the requirements of the individual units.

After the input data is read and printed, the main program calls the subroutine COST, which performs the necessary computations and returns the results to the main program for printing. Some of the more important output reports that are printed by the main routine are

- the distribution of shortages,
- statistics of generation by units of the system,
- the load–duration curve of demand,
- the combined load–duration curve of demand plus maintenance,
- the combined load–duration curve of demand plus maintenance plus breakdowns.

B Subroutine COST

This is the central subroutine of the program. Its purpose is to call other subroutines and to perform the computations of the cost of operating the system.

The *subroutines that are called* by subroutine COST are

- subroutine WEEK—to treat each new week,
- subroutine TAKALA—to perform random number generation that decides whether a particular unit is nonoperative due to a breakdown,
- subroutine SHORT—to compute the distribution of shortages in cases where it is impossible to satisfy all of the demand (i.e., when the demand is greater than the effective generating capacity),
- subroutine CUMLDC—this subroutine is called three times: to compute the load–duration curve of demand only; to compute the combined load–duration curve of demand plus maintenance; and to compute the combined load–duration curve of demand plus maintenance plus breakdowns.

The computation of operating costs is performed by means of three loops.

In the first loop the subroutine establishes which generating units are to operate during the week. The operating units are the ones that are neither in maintenance nor broken down when called for generation. (If the peak demand of the week is low, there may exist units which are in good working order but which need not be operated because of insufficient demand.) The basic assumption used in scheduling the units for generation is that of the merit order. That is, the scheduling is performed in the sequence of the efficiency of the units, so that the most efficient unit (from the point of view of cost of fuel per kilowatt-hour) operates first, the next most efficient operates next, and so forth until the least efficient unit.

The second loop deals with the problem of the minimum level of operation. Some operating units are prohibited from operating at a load that is lower than some prespecified level, which we will call the *minimum level of operation*. In certain units this minimum reaches 40% of the maximum generating capacity of the unit. This restriction on operation forces us first to assign to operation the minimum level of operation of all the units, and only afterward to assign the remaining generating capacity of the units according to the merit order principle.

The basic assumption regarding minimum level of operation is that the sum of the minimum levels of operation of all the available units during the week may not be greater than the base load. Otherwise, a situation will arise in which some unit will be required, for some period of time, to operate at a lower level than its minimum level of operation.

The third loop assigns to operation the remaining operating capacity of all operating units according to their merit order.

While scheduling the units for operation, subroutine COST computes the energy generated according to the weekly load–duration curve kept in memory. Multiplication of the amount of energy by the fuel costs, according to region and type of plant, yields the cost of weekly operation.

The weekly costs are accumulated in order to compute the total annual cost. These annual costs, multiplied by suitable discount factors, are accumulated for computing the total discounted cost of the investigated alternative.

C Subroutine SHORT

The purpose of this subroutine is to compute the distribution of shortages in the cases where the system has insufficient capacity for satisfying the demand. It follows that this subroutine is entered only in the case where demand for electricity exceeds the production capability.

The principle for computing the distribution of shortages is as follows. We divide the difference between the peak demand and the amount of electricity supplied (i.e., the magnitude of shortage) into small units of 10 MW each. In this fashion the top portion of the weekly load–duration curve is divided into trapezoids. For each trapezoid the subroutine computes the number of hours of shortage and the unsupplied energy (which equals the area of the trapezoid). The number of hours of shortage is accumulated into a vector for computing the distribution of shortages. Similarly, the areas of the trapezoids are accumulated for computing the total unsupplied energy.

D Subroutine WEEK

This subroutine is entered at the beginning of the processing of each new week. (The program processes a maximum of 520 weeks.) The subroutine first initializes the counters that accumulate the week numbers (YRCTR,

WKCTR). It then computes the weekly load–duration curve based on the typical load–duration curves, as follows. The program first identifies, by means of the vector TWEEK, which of the typical load–duration curves (described in TYPE) are to be related to the week in question. (There are three different typical load–duration curves.) Then, on the basis of the expected average demand for that week (taken from vector DEMAND), it computes the load–duration curve of the week.

The load–duration curve, which is a continuous curve, is approximated by subroutine WEEK by means of straight-line segments. The point of intersection of each two segments that describe the load–duration curve is referred to as a *vertex point*. For each such point (there are five in all) the subroutine WEEK computes the ordinate (megawatts) and the abscissa (hours), and also the slope of the segment that extends from the point downward and to the right and the intercept of the segment.

The main results of the computations are stored in the matrix LDC of dimension (5,5), and are transferred back into subroutine COST for the purpose of subsequent computations.

E Subroutine MERTSB

The purpose of this subroutine is to update the merit order vector (MERIT), i.e., the vector that establishes the order of relative efficiency of the generating units in the system.

This subroutine is executed each time that a new unit is introduced into the system. The procedure for updating the merit order is as follows:

- If the generating unit entering is a nuclear unit, then it is the most efficient of all of the units operating in the system.
- If the generating unit entering is a conventional unit, then it is more efficient than any of the conventional units and gas turbines that are operating in the system, but it is less efficient than the nuclear units, if there exist any in the system.
- If the generating unit entering is a gas turbine, then it is more efficient than any of the gas turbines that are operating in the system, but it is less efficient than the conventional units and the nuclear units.

We emphasize that the merit order of the units that already exist in the system is input data to the program; the subroutine updates this vector only for the new units that enter into the system.

F Subroutine CUMLDC

This subroutine computes the load–duration curves and it is called by subroutine COST.

At the completion of the weekly computations, the call to subroutine CUMLDC is performed three times: the first time to compute the load–duration curve of demand only (LOADA); the second time to compute the load–duration curve that includes demand and maintenance (LOADB); and the third time to compute the combined load–duration curve that includes demand, maintenance, and breakdowns (LOADC).

In order to compute the combined load–duration curves we divide the vertical axis of the weekly load–duration curve (the megawatts axis) into segments of 10 MW per segment, and we accumulate the hours (on the horizontal axis) of each of these segments.

G Subroutines CONVRT and DATE

These subroutines are used for translating dates (such as the date that a new unit is being introduced) into week numbers relative to the base date.

H Subroutine TAKALA

This subroutine generates random breakdowns for each generating unit for each week. If we denote the proportion of time spent by unit i of the system at a forced breakdown by means of q_i, then the probability that the generating unit will not break down in any unit of time is $p_i = 1 - q_i$.

The process of random generation of breakdown is performed as follows.

A random number is generated in the range between 0 and 1 by subroutine RANDU. We will denote this number by R.

If $R \leq q_i$, then the unit is broken down (forced out).

If $R > q_i$, then the unit is in operation.

This process of generating breakdowns is performed only for the units that are potentially in operation, i.e., units that are not in maintenance and units such that the demand is sufficient to warrant their operation.

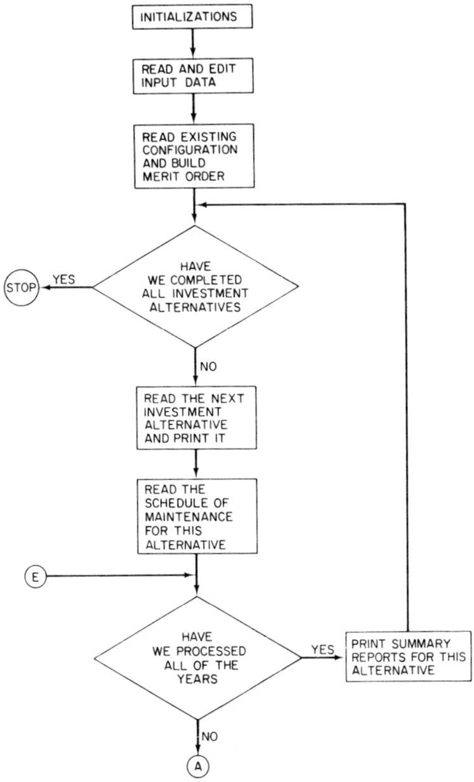
INITIALIZATIONS
READ AND EDIT INPUT DATA
READ EXISTING CONFIGURATION AND BUILD MERIT ORDER
HAVE WE COMPLETED ALL INVESTMENT ALTERNATIVES
YES
STOP
NO
READ THE NEXT INVESTMENT ALTERNATIVE AND PRINT IT
READ THE SCHEDULE OF MAINTENANCE FOR THIS ALTERNATIVE
E
HAVE WE PROCESSED ALL OF THE YEARS
YES
PRINT SUMMARY REPORTS FOR THIS ALTERNATIVE
NO
A

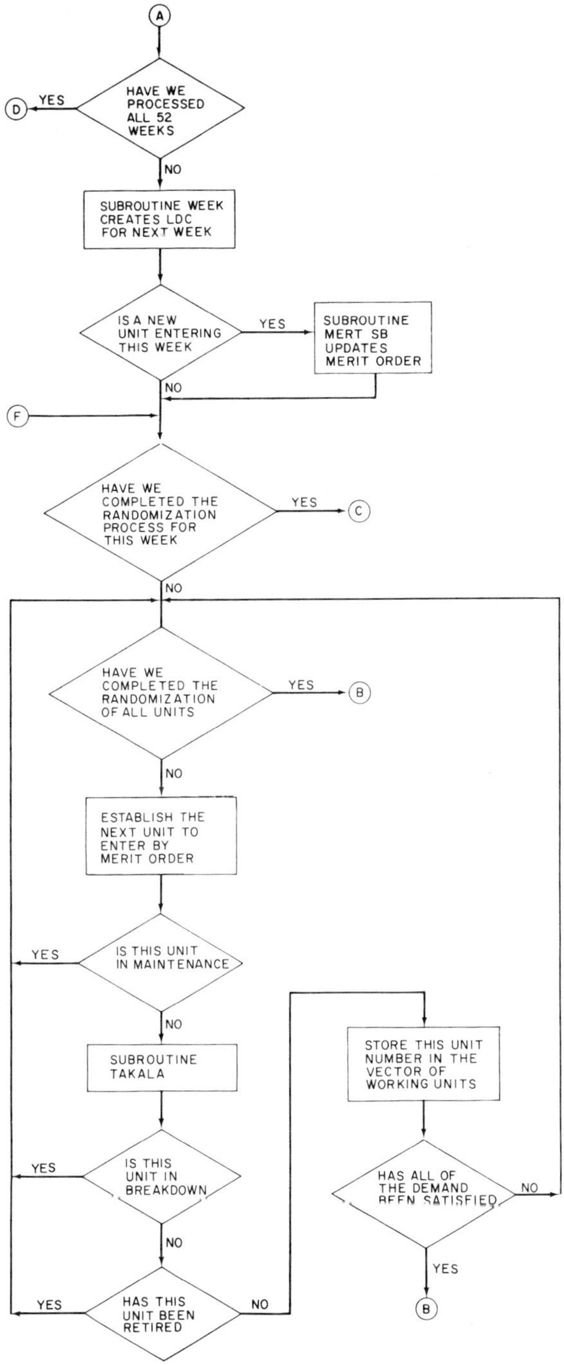
A
HAVE WE PROCESSED ALL 52 WEEKS
YES
D
NO
SUBROUTINE WEEK CREATES LDC FOR NEXT WEEK
IS A NEW UNIT ENTERING THIS WEEK
YES
SUBROUTINE MERT SB UPDATES MERIT ORDER
NO
F
HAVE WE COMPLETED THE RANDOMIZATION PROCESS FOR THIS WEEK
YES
C
NO
HAVE WE COMPLETED THE RANDOMIZATION OF ALL UNITS
YES
B
NO
ESTABLISH THE NEXT UNIT TO ENTER BY MERIT ORDER
IS THIS UNIT IN MAINTENANCE
YES
NO
SUBROUTINE TAKALA
IS THIS UNIT IN BREAKDOWN
YES
NO
HAS THIS UNIT BEEN RETIRED
YES
NO
STORE THIS UNIT NUMBER IN THE VECTOR OF WORKING UNITS
HAS ALL OF THE DEMAND BEEN SATISFIED
NO
YES
B

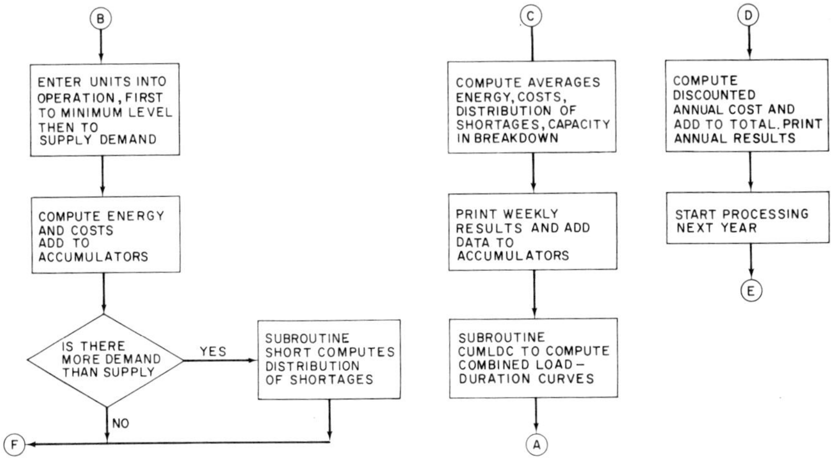
B
ENTER UNITS INTO OPERATION, FIRST TO MINIMUM LEVEL THEN TO SUPPLY DEMAND
COMPUTE ENERGY AND COSTS ADD TO ACCUMULATORS
IS THERE MORE DEMAND THAN SUPPLY
YES
NO
SUBROUTINE SHORT COMPUTES DISTRIBUTION OF SHORTAGES
F
C
COMPUTE AVERAGES ENERGY, COSTS, DISTRIBUTION OF SHORTAGES, CAPACITY IN BREAKDOWN
PRINT WEEKLY RESULTS AND ADD DATA TO ACCUMULATORS
SUBROUTINE CUMLDC TO COMPUTE COMBINED LOAD – DURATION CURVES
A
D
COMPUTE DISCOUNTED ANNUAL COST AND ADD TO TOTAL. PRINT ANNUAL RESULTS
START PROCESSING NEXT YEAR
E

List of Symbols and Abbreviations

a	unit's availability
c	operating costs
c'	marginal operating costs
E	energy
ELDC	equivalent load–duration curve (sometimes referred to as CLDC—combined load–duration curve)
F	fixed costs
$f(p)$	production function
FOR	forced outages rate
LDC	load–duration curve
LOEP	loss-of-energy probability
LOLP	loss-of-load probability
p	capacity, demand, load (general)
p_{C}	combined demand
p_{D}	customers' demand (deterministic component)
$p_{\mathrm{D\,max}}$	peak customers' demand
p_{F}	capacity under forced outage
p_{ins}	the installed capacity
p_{L}	the load
p_{M}	the capacity under maintenance
p_{P}	the capacity of unit added to the system
p_{R}	the random component of customers' demand
$P(f)$	probability function
q	outage probability
S	set of all events of in and out units
$\bar{S}$	new set after adding a unit
t	duration
$T(p)$	load–duration curve (hours)

$t(p)$	load–duration curve (fraction of time)
$\bar{t}_i(p)$	combined load–duration curve of first i units
$\bar{T}(p)$	combined load–duration curve of the system (hours)
$\bar{t}(p)$	combined load–duration curve of the system (fraction of time)
$\bar{\bar{t}}(p)$, $\hat{t}(p)$, $t(p)$	various combined load–duration curves under different considerations
U	unit loading
V	variable costs

References

1. Anderson, D. (1972). "Models for Determining Least Cost Investments in Electricity Supply," *Bell Journal of Economics and Management Science* 3: 267–299.

2. Balasko, Y. (1974) "On Designing Public Utilities Tariffs with Applications to Electricity," Manuscript, Environmental Defence fund (EDF), New York.

3. Baleriaux, H., Jamoulle, E., and Linard de Guertechin, Fr. (1967). "Simulation de l'exploitation d'un parc de machines thermique de production d'electricité couple à des stations de pampage," *Review E* (*SRBE*) 5: 225–245.

4. Berrie, T. W. (1968). "The Economics of System Planning in Bulk Electricity Supply," in *Public Enterprise*, edited by R. Turvey, Penguin Modern Economics, Penguin, London.

5. Bierman, H., Jr., and Hass, J. E. (1975). "Inflation, Equity, Efficiency, and the Regulatory Pricing of Electricity," *Public Policy* 22: 299–314.

6. Billinton, R. (1970). *Power System Reliability Evaluation*, Gordon and Breach, New York.

7. Billinton, R. (1972). "Bibliography on the Application of Probability Methods in Power System Reliability Evaluation," *IEEE Transactions* PAS-91: 649–660.

8. Billinton, R., and Krasnodebski, J. (1973). "Practical Application of Reliability and Maintainability Concepts to Generating Station Design," *IEEE Transactions* PAS-92: 1814–1824.

9. Booth, R. R. (1972). "Power System Simulation Model Based on Probability Analysis," *IEEE Transactions* PAS-91: 62–69.

10. Crew, M. A., and Kleindorfer, P. R. (1976). "Peak Load Pricing with a Diverse Technology," *The Bell Journal of Economics* 7: 207–231.

11. Edison Electric Institute (1977). "Statistical Yearbook of the Electric Utility Industry for 1976," New York.

12. Elgerd, O. I. (1971). *Electric Energy Systems Theory: An Introduction*, McGraw–Hill, New York.

13. Federal Energy Administration (1975). "A Report on Improving on Productivity of Electric Power Plants." Washington, D. C.

14. Galloway, C. D., Garver, L. L., Kirchmayer, L. K., and Wood, A. J. (1969). "Generation Transmission Planning and Economic Evaluation," Proceedings of the Third Power Systems Computation Conference, Rome.

15. Jenkins, R. T., and Joy, D. S. (1974). "WIEN Automatic System Planning Package (WASP)—An Electric Utility Optimal Generation Expansion Planning Computer Code," Report ORNL-4945, Oak Ridge National Laboratory, July 1974, Oak Ridge, Tennessee.

16. Joskow, P. L. (1975). "Marginal Cost Pricing of Electricity," Testimony before the Public Service Commission of New York, National Economic Research Associates (NERA), New York.

17. Joskow, P. L. (1976). "Contributions to the Theory of Marginal Cost Pricing," *The Bell Journal of Economics* 7: 197–206.

18. Joskow, P. L., and Baughman, M. I. (1976). "The Future of the U.S. Nuclear Energy Industry," *The Bell Journal of Economics* 7: 3–32.

19. Joy, D. S., and Jenkins, R. T. (1974). "A Probabilistic Model for Estimating the Operating Cost of an Electric Power Generating System," Papers presented at the Nuclear Utilities Planning Methods Symposium, Oak Ridge National Laboratory, January 1974, Oak Ridge, Tennessee.

20. Kahn, A. E. (1970). *The Economics of Regulation: Principles and Institutions*, Vol. 1, Wiley, New York.

21. Kaufman, A. (1975). "Reliability Criteria—A Cost Benefit Analysis," Report 75–9, New York State Department of Public Service, New York.

22. Mahoney, L. T. Jr. (1975). "Cost Analysis for Use in Peak Load Pricing," Testimony before the Public Service Commission of New York, National Economic Research Associates (NERA), New York.

23. Marsh, W. D., Moisan, R. W., and Murrell, T. C. (1974). "Perspectives on the Design and Application of Generation Planning Program," Papers presented at the Nuclear Utilities Planning Methods Symposium, Oak Ridge National Laboratory, January 1974, Oak Ridge, Tennessee.

24. Nelson, J. R. (1964). *Marginal Cost Pricing in Practice*, Prentice–Hall, Englewood Cliffs, New Jersey.

25. Phillips, D., Jenkin, F. P., Pritchard, J. A. T., and Rybicki, K. (1969). "A Mathematical Model for Determining Generation Plant Mix," Third Power Systems Computation Conference (PSCC), Rome.

26. Ringlee, R. J., Hall, J. D., and Wood, A. J. (1968). "Frequency and Duration Methods for Power System Reliability Calculations: I—Generation System Model," *IEEE Transactions* PAS-87: 1787–1796.

27. Rogers, J. S. (1974). "The Cost of Forced Outages—A micro-Economics Approach," Paper C 74 145–9 presented at the IEEE PES Meeting, January 1974, New York.

28. Sager, M. A., and Wood, A. J. (1973). "Power System Production Cost Calculation—Sample Studies Recognizing Forced Outages," *IEEE Transactions* PAS-92: 154–158.

29. Sager, M. A., Ringlee, R. J., and Wood, A. J. (1972). "A New Generation Production Cost Program to Recognize Forced Outages," *IEEE Transactions* PAS-91: 2114–2124.

30. Schofield, G. H., and Marsh, W. D. (1970). "The Effect of Reliability on Utility Financial Results," *Public Utilities Fortnightly* 86: 15–20.

31. Streiter, S. H. (1975). "Peak Load Costing and Pricing Methodology," Testimony before the Public Service Commission of New York, National Economic Research Associates (NERA), New York.

32. Telson, M. L. (1975). "The Economics of Alternative Levels of Reliability for Electric Power Generation System," *The Bell Journal of Economics* 6: 679–694.

33. Turvey, R. (1968). "*Optimal Pricing and Investment in Electricity Supply*," MIT Press, Cambridge, Massachusetts.

34. U.S. Federal Power Commission (1971). *The 1970 National Power Survey*, U.S. Government Printing Office, Washington, D. C.

35. Vardi, J., Zahavi, J., and Avi-Itzhak, B. (1976). "Reliability and Sensitivity Analysis in Power Systems," Paper A 76 346–7 presented at the IEEE PES Summer Meeting, Portland, Oregon, July, 1976.

36. Vardi, J., Zahavi, J., and Avi-Itzhak, B. (1977). "Variable Load Pricing in the Face of Loss of Load Probability," *The Bell Journal of Economics* 8: 270–288.

37. Vardi, J., Zahavi, J., and Avi-Itzhak, B. (1977). "The Combined Load Duration Curve and Its Derivation," *IEEE Transactions* PAS-96: 978–983.

38. Wenders, J. T. (1976). "Peak Load Pricing in the Electric Utility Industry." *The Bell Journal of Economics* 7: 232–241.

39. Zahavi, J. (1975). "An Approach to Solution Analysis With An Application to Generation Capacity Expansion of Power Systems," Technical Report No. 272, Department of Operations Research, Cornell University, Ithaca, New York.

40. Zahavi, J., Eisenberg, L. (1975). "Economic Environmental Power Dispatch," *IEEE Transactions*, SMC-5: 485–489.

41. Zahavi, J., Vardi, J., and Avi-Itzhak, B. (1977). "Operating Cost Calculation of an Electric Power Generating System Under Incremental Loading Procedure," *IEEE Transactions* PAS-96: 285–292.

42. Zahavi, J., Vardi, J., and Avi-Itzhak, B. (1978). "A Unified Approach to Support Planning Decisions in Electric Power Generating Systems," *Studies in the Management Sciences*, Vol. 10, pp. 197–160, North-Helland, New York.

Index